QH
88
E95

Evolution of desert biota.

P9-EKE-744

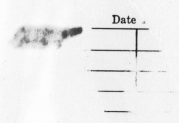

Date

Evolution of Desert Biota

Evolution of Desert Biota

Edited by David W. Goodall

University of Texas Press Austin & London

Publication of this book was financed in part by
the Desert Biome and the Structure of Ecosystems programs of
the U.S. participation in the International
Biological Program.

Library of Congress Cataloging in Publication Data
Main entry under title:

Evolution of desert biota.

   Proceedings of a symposium held during the First
International Congress of Systematic and Evolution-
ary Biology which took place in Boulder, Colo.,
during August, 1973.
   Bibliography: p.
   Includes index.
   1. Desert biology—Congresses. 2. Evolution—
Congresses. I. Goodall, David W., 1914–
II. International Congress of Systematic and Evolu-
tionary Biology, 1st, Boulder, Colo., 1973.
QH88.E95   575'.00915'4   75-16071
ISBN 0-292-72015-7

Copyright © 1976 by the University of Texas Press

All rights reserved

Printed in the United States of America

Contents

Q H
8 8
E 95

76 03595

Evolution of Desert Biota

1. Introduction   David W. Goodall

In the broad sense, "deserts" include all those areas of the earth's surface whose biological potentialities are severely limited by lack of water. If one takes them as coextensive with the arid and semiarid zones of Meigs's classification, they occupy almost one-quarter of the terrestrial surface of the globe. Though the largest arid areas are to be found in Africa and Asia, Australia has the largest proportion of its area in this category. Smaller desert areas occur in North and South America; Antarctica has cold deserts; and the only continent virtually without deserts is Europe.

When life emerged in the waters of the primeval world, it could hardly have been predicted that the progeny of these first organisms would extend their occupancy even to the deserts. Regions more different in character from the origin and natural home of life would be hard to imagine. Protoplasm is based on water, rooted in water. Some three-quarters of the mass of active protoplasm is water; the biochemical reactions underlying all its activities take place in water and depend on the special properties of water for the complex mechanisms of enzymatic and ionic controls which integrate the activity of cell and organisms into a cybernetic whole. It is, accordingly, remarkable that organisms were able to adapt themselves to environments in which water supplies were usually scanty, often almost nonexistent, and always unpredictable.

The first inhabitants of the deserts were presumably opportunistic. On the margins of larger bodies of water were areas which were alternately wetted and dried for longer or shorter periods. Organisms living there acquired the possibility of surviving the dry periods by drying out and becoming inactive until rewetted, at which time their activity resumed where it had left off. While in the dry state, these organisms

—initially, doubtless, Protista—were easily moved by air currents and thus could colonize other bodies of water. Among them were the very temporary pools formed by the occasional rainstorms in desert areas. Thus the deserts came to be inhabited by organisms whose ability to dry and remoisten without loss of vitality enabled them to take advantage of the short periods during which limited areas of the deserts deviate from their normally arid state.

Yet other organisms doubtless—the blue green algae among them —similarly took advantage of the much shorter periods, amounting perhaps to an hour at a time, during which the surface of the desert was moistened by dew, and photosynthesis was possible a few minutes before and after sunrise to an organism which could readily change its state of hydration.

In the main, though, colonization of the deserts had to wait until colonization of other terrestrial environments was well advanced. For most groups of organisms, the humid environments on land presented less of a challenge in the transition from aquatic life than did the deserts. By the time arthropods and annelids, mollusks and vertebrates, fungi and higher plants had adapted to the humid terrestrial environments, they were poised on the springboard where they could collect themselves for the ultimate leap into the deserts. And this leap was made successfully and repeatedly. Few of the major groups of organisms that were able to adapt to life on land did not also contrive to colonize the deserts.

Some, like the arthropods and annual plants, had an adaptational mechanism—an inactive stage of the life cycle highly resistant to desiccation—almost made to order to match opportunistically the episodic character of the desert environment. For others the transition was more difficult: for mammals, whose excretory mechanism assumes the availability of liquid water; for perennial plants, whose photosynthetic mechanism normally carries the penalty of water loss concurrent with carbon dioxide intake. But the evolutionary process surmounted these difficulties; and the deserts are now inhabited by a range of organisms which, though somewhat inferior to that of more favored environments, bears testimony to the inventiveness and success of evolution in filling niches and in creating diversity.

The most important modifications and adaptations needed for life in the deserts are concerned with the dryness of the environment there.

But an important feature of most desert environments is also their unpredictability. Precipitation has a higher coefficient of variability, on any time scale, than in other climatic types, with the consequence that desert organisms may have to face floods as well as long and highly variable periods of drought. The range of temperatures may also be extreme—both diurnal and seasonal. Under the high radiation of the subtropical deserts, the soil surface may reach a temperature which few organisms can survive; and, in the cold deserts of the great Asian land mass, extremely low winter temperatures are recorded. Sand and dust storms made possible by the poor stability of the surface soil are also among the environmental hazards to which desert organisms must become adapted.

Like other climatic zones, the deserts have not been stable in relation to the land masses of the world. Continental drift, tectonic movements, and changes in the earth's rotation and in the extent of the polar icecaps have led to secular changes in the area and distribution of arid land surfaces. But, unlike other climatic zones, the arid lands have probably always been fragmented—constituting a number of discrete areas separated from one another by zones of quite different climate. The evolutionary process has gone on largely independently in these separate areas, often starting from different initial material, with the consequence that the desert biota is highly regional. Elements in common between the different main desert areas are few, and, as between continents or subcontinents, there is a high degree of endemism. The smaller desert areas of the world are the equivalent of islands in an ocean of more humid environments.

These are among the problems to be considered in the present volume. It reports the proceedings of a symposium which was held on August 10, 1973, at Boulder, Colorado, as part of the First International Congress of Systematic and Evolutionary Biology.

2. The Origin and Floristic Affinities of the South American Temperate Desert and Semidesert Regions   Otto T. Solbrig

Introduction

In this paper I will attempt to summarize the existent evidence regarding the floristic relations of the desert and semidesert regions of temperate South America and to explain how these affinities came to exist.

More than half of the surface of South America south of the Tropic of Capricorn can be classed as semidesert or desert. In this area lie some of the richest mineral deposits of the continent. These regions consequently are important from the standpoint of human economy. From a more theoretical point, desert environments are credited with stimulating rapid evolution (Stebbins, 1952; Axelrod, 1967) and, further, present some of the most interesting and easy-to-study adaptations in plants and animals.

Although, at present, direct evidence regarding the evolution of desert vegetation in South America is still meager, enough data have accumulated to make some hypotheses. It is hoped this will stimulate more research in the field of plant micropaleontology in temperate South America. Such research in northern South America has advanced our knowledge immensely (Van der Hammen, 1966), and high rewards await the investigator who searches this area in the temperate regions of the continent.

The Problem

If a climatic map of temperate South America is compared with a phytogeographic map of the same region drawn on a physiognomic basis and with one drawn on floristic lines, it will be seen that they do not coincide. Furthermore, if the premise (not proven but generally held to be true) is accepted that the physical environment is the determinant of the structure of the ecosystem and that, as the physical environment (be it climate, physiography, or both) changes, the structure of the vegetation will also change, then an explanation for the discrepancy between climatic and phytogeographic maps has to be provided. Alternative explanations to solve the paradox are (1) the premise on which they are based is entirely or partly wrong; (2) our knowledge is incomplete; or (3) the discrepancies can be explained on the basis of the historical events of the past. It is undoubtedly true that floristic and paleobotanical knowledge of South American deserts is incomplete and that much more work is needed. However, I will proceed under the assumption that a sufficient minimum of information is available. I also feel that our present insights are sufficient to accept the premise that the ecosystem is the result of the interaction between the physical environment and the biota. I shall therefore try to find in the events of the past the answer for the discrepancy.

I shall first describe the semidesert regions of South America and their vegetation, followed by a brief discussion of Tertiary and Pleistocene events. I shall then look at the floristic connections between the regions and the distributional patterns of the dominant elements of the area under study. From this composite picture I shall try to provide a coherent working hypothesis to explain the origin and floristic affinities of the desert and semidesert regions of temperate South America.

Theory

Biogeographical hypotheses such as the ones that will be made further on in this paper are based on certain theoretical assumptions. In most cases, however, these assumptions are not made explicit; consequently, the reader who disagrees with the author is not always certain whether he disagrees with the interpretation of the evidence or with the assumptions made. This has led to many futile controversies. The fundamental assumptions that will be made here follow from the general theory of evolution by natural selection, the theory of speciation, and the theory of geological uniformitarianism.

The first assumption is that a continuous distributional range reflects an environment favorable to the plant, that is, an environment where it can compete successfully. Since the set of conditions (physical, climatical, and biological) where the plant can compete successfully (the realized niche) bounds a limited portion of ecological space, it will be further assumed that the range of a species indicates that conditions over that range do not differ greatly in comparison with the set of all possible conditions that can be given. It will be further assumed that each species is unique in its fundamental and realized niche (defined as the hyperspace bounded by all the ecological parameters to which the species is adapted or over which it is able to compete successfully). Consequently, no species will occupy exactly the same geographical range, and, as a corollary, some species will be able to grow over a wide array of conditions and others over a very limited one.

When the vegetation of a large region, such as a continent, is mapped, it is found that the distributional ranges of species are not independent but that ranges of certain species tend to group themselves even though identical ranges are not necessarily encountered. This allows the phytogeographer to classify the vegetation. It will be assumed that, when neighboring geographical areas do not differ greatly in their present physical environment or in their climate but differ in their flora, the reason for the difference is a historical one reflecting different evolutionary histories in these floras and requiring an explanation.

Disjunctions are common occurrences in the ranges of species. In a strict sense, all ranges are disjunct since a continuous cover of a species over an extensive area is seldom encountered. However, when similar major disjunctions are found in the ranges of many species whose ranges are correlated, the disjunction has biogeographical significance. Unless there is evidence to the contrary, an ancient continuous range will be assumed in such instances, one that was disrupted at a later date by some identifiable event, either geological or climatological.

It will also be assumed that the atmospheric circulation and the basic meteorological phenomena in the past were essentially similar to those encountered today, unless there is positive evidence to the contrary. Further, it will be assumed that the climatic tolerances of a living species were the same in the past as they are today. Finally, it will be assumed that the spectrum of life forms that today signify a rain forest, a subtropical forest, a semidesert, and so on, had the same meaning in the past too, implying with it that the basic processes of carbon gain and water economy have been essentially identical at least since the origin of the angiosperms.

From these assumptions a coherent theory can be developed to reconstruct the past (Good, 1953; Darlington, 1957, 1965). No general assumptions about age and area will be made, however, because they are inconsistent with speciation theory (Stebbins, 1950; Mayr, 1963). In special cases when there is some evidence that a particular group is phylogenetically primitive, the assumption will be made that it is also geologically old. Such an assumption is not very strong and will be used only to support more robust evidence.

The Semidesert Regions of South America

In temperate South America we can recognize five broad phytogeographical regions that can be classed as "desert" or "semidesert" regions. They are the Monte (Haumann, 1947; Morello, 1958), the Patagonian Steppe (Cabrera, 1947), the Prepuna (Cabrera, 1971), and the Puna (Cabrera, 1958) in Argentina, and the Pacific Coastal Desert in Chile and Peru (Goodspeed, 1945; Ferreyra, 1960). In addition, three other regions—the Matorral or "Mediterranean" region in Chile (Mooney and Dunn, 1970) and the Chaco and the Espinal in Argentina (Fiebrig, 1933; Cabrera, 1953, 1971), although not semideserts, are characterized by an extensive dry season. Finally, the high mountain vegetation of the Andes shows adaptations to drought winters (fig. 2-1).

The Monte

The Monte (Lorentz, 1876; Haumann, 1947; Cabrera, 1953; Morello, 1958; Solbrig, 1972, 1973) is a phytogeographical province that extends from lat. 24°35′ S to lat. 44°20′ S and from long. 62°54′ W on the Atlantic coast to long. 69°50′ W at the foothills of the Andes (fig. 2-1).

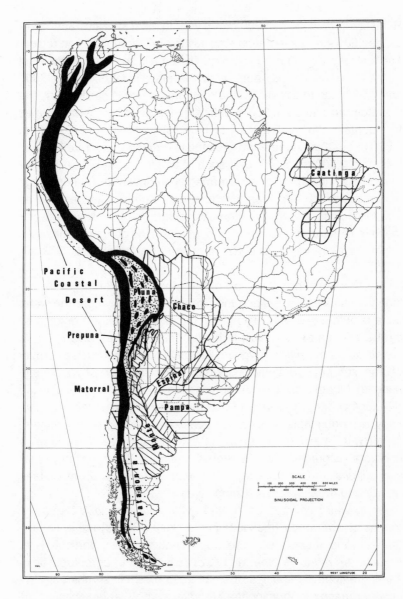

Fig. 2-1. *Geographical limits of the phytogeographical provinces of the Andean Dominion (stippled) and of the Chaco Dominion (various hatchings) according to Cabrera (1971). The high cordillera vegetation is indicated in solid black. Goode Base Map, copyright by The University of Chicago, Department of Geography.*

Rains average less than 200 mm a year in most localities and never exceed 600 mm; evaporation exceeds rainfall throughout the region. The rain falls in spring and summer. The area is bordered on the west by the Cordillera de los Andes, which varies in height between 5,000 and 7,000 m in this area. On the north the region is bordered by the high Bolivian plateau (3,000–5,000 m high) and on the east by a series of mountain chains (Sierras Pampeanas) that vary in height from 3,000 to 5,000 m in the north (Aconquija, Famatina, and Velazco) to less than 1,000 m (Sierra de Hauca Mahuida) in the south. Physiographically, the northern part is formed by a continuous barrier of high mountains which becomes less important farther south as well as lower in height. The Monte vegetation occupies the valleys between these mountains as a discontinuous phase in the northern region and a more or less continuous phase from approximately lat. 32° S southward.

The predominant vegetation of the Monte is a xerophytic scrubland with small forests along the rivers or in areas where the water table is quite superficial. The predominant community is dominated by the species of the creosote bush or *jarilla* (*Larrea divaricata*, *L. cuneifolia*, and *L. nitida* [Zygophyllaceae]) associated with a number of other xerophytic or aphyllous shrubs: *Condalia microphylla* (Rhamnaceae), *Monttea aphylla* (Scrophulariaceae), *Bougainvillea spinosa* (Nyctaginaceae), *Geoffroea decorticans* (Leguminosae), *Cassia aphylla* (Leguminosae), *Bulnesia schickendanzii* (Zygophyllaceae), *B. retama*, *Atamisquea emarginata* (Capparidaceae), *Zuccagnia punctata* (Leguminosae), *Gochnatia glutinosa* (Compositae), *Proustia cuneifolia* (Compositae), *Flourensia polyclada* (Compositae), and *Chuquiraga erinacea* (Compositae).

Along water courses or in areas with a superficial water table, forests of *algarrobos* (mesquite in the United States) are observed, that is, various species of *Prosopis* (Leguminosae), particularly *P. flexuosa*, *P. nigra*, *P. alba*, and *P. chilensis*. Other phreatophytic or semiphreatophytic species of small trees or small shrubs are *Cercidium praecox* (Leguminosae), *Acacia aroma* (Leguminosae), and *Salix humboldtiana* (Salicaceae).

Herbaceous elements are not common. There is a flora of summer annuals formed principally by grasses.

The Patagonian Steppe

The Patagonian Steppe (Cabrera, 1947, 1953, 1971; Soriano, 1950, 1956) is limited on its eastern and southern borders by the Atlantic Ocean and the Strait of Magellan. On the west it borders quite abruptly with the *Nothofagus* forest; the exact limits, although easy to determine, have not yet been mapped precisely (Dimitri, 1972). On the north it borders with the Monte along an irregular line that goes from Chos Malal in the state of Neuquen in the west to a point on the Atlantic coast near Rawson in the state of Chubut (Soriano, 1949). In addition, a tongue of Patagonian Steppe extends north from Chubut to Mendoza (Cabrera, 1947; Böcher, Hjerting, and Rahn, 1963). Physiognomically the region consists of a series of broad tablelands of increasing altitude as one moves from east to west, reaching to about 1,500 m at the foot of the cordillera. The soil is sandy or rocky, formed by a mixture of windblown cordilleran detritus as well as *in situ* eroded basaltic rocks, the result of ancient volcanism.

The climate is cold temperate with cold summers and relatively mild winters. Summer means vary from 21°C in the north to 12°C in the south (summer mean maxima vary from 30°C to 18°C) with winter means from 8°C in the north to 0°C in the south (winter mean minima 1.5°C to −3°C). Rainfall is very low, averaging less than 200 mm in all the Patagonian territory with the exception of the south and west borders where the effect of the cordilleran rainfall is felt. The little rainfall is fairly well distributed throughout the year with a slight increase during winter months.

The Patagonian Steppe is the result of the rain-shadow effect of the southern cordillera in elevating and drying the moist westerly winds from the Pacific. Consequently the region not only is devoid of rains but also is subjected to a steady westerly wind of fair intensity that has a tremendous drying effect. The few rains that occur are the result of occasional eruptions of the Antarctic polar air mass from the south interrupting the steady flow of the westerlies.

The dominant vegetation is a low scrubland or else a vegetation of low cushion plants. In some areas xerophytic bunch grasses are also common. Among the low (less than 1 m) xerophytic shrubs and cushion plants, the *neneo*, *Mulinum spinosum* (Umbelliferae), is the dominant form in the northwestern part, while *Chuquiraga avellanedae* (Compositae) and *Nassauvia glomerulosa* (Compositae) are dominant over extensive areas in central Patagonia. Other important shrubs are *Trevoa patagonica* (Rhamnaceae), *Adesmia campestris* (Compositae), *Colliguaja integerrima* (Euphorbiaceae), *Nardophyllum obtusifolium* (Compositae), and *Nassauvia axillaris*. Among the grasses are *Stipa humilis*, *S. neaei*, *S. speciosa*, *Poa huecu*, *P. ligularis*, *Festuca argentina*, *F. gracillima*, *Bromus macranthus*, *Hordeum comosus*, and *Agropyron fuegianum*.

The Puna

The Puna (Weberbauer, 1945; Cabrera, 1953, 1958, 1971) is situated in the northwestern part of Argentina, western and central Bolivia, and southern Peru. It is a very high plateau, the result of the uplift of an enormous block of an old peneplane, which started to lift in the Miocene but mainly rose during the Pliocene and the Pleistocene to a mean elevation of 3,400–3,800 m. The Puna is bordered on the east by the Cordillera Real and on the west by the Cordillera de los Andes that rises to 5,000–6,000 m; the plateau is peppered by a number of volcanoes that rise 1,000–1,500 m over the surface of the Puna.

The soils of the Puna are in general immature, sandy to rocky, and very poor in organic matter (Cabrera, 1958). The area has a number of closed basins, and high mountain lakes and marshes are frequent.

The climate of the Puna is cold and dry with values for minimum and maximum temperatures not too different from Patagonia but with the very significant difference that the daily temperature amplitude is very great (values of over 20°C are common) and the difference between summer and winter very slight. The precipitation is very irregular over the area of the Puna, varying from a high of 800 mm in the northeast corner of Bolivia to 100 mm/year on the southwest border in Argentina. The southern Puna is undoubtedly a semidesert region, but the northern part is more of a high alpine plateau, where the limitations to plant growth are given more by temperature than by rainfall.

The typical vegetation of the Puna is a low, xerophytic scrubland formed by shrubs one-half to one meter tall. In some areas a grassy steppe community is found, and in low areas communities of high mountain marshes are found.

Among the shrubby species we find *Fabiana densa* (Solanaceae), *Psila boliviensis* (Compositae), *Adesmia horridiuscula* (Leguminosae), *A. spinossisima, Junellia seriphioides* (Verbenaceae), *Nardophyllum armatum* (Compositae), and *Acantholippia hastatula* (Verbenaceae). Only one tree, *Polylepis tomentella* (Rosaceae), grows in the Puna, strangely enough only at altitudes of over 4,000 m. Another woody element is *Prosopis ferox*, a small tree or large shrub. Among the grasses are *Bouteloua simplex, Muhlenbergia fastigiata, Stipa leptostachya, Pennisetum chilense*, and *Festuca scirpifolia*. Cactaceae are not very frequent in general, but we find locally abundant *Opuntia atacamensis, Oreocerus trollii, Parodia schroebsia*, and *Trichocereus poco*.

Although physically the Puna ends at about lat. 30° S, Puna vegetation extends on the eastern slope of the Andes to lat. 35° S, where it merges into Patagonian Steppe vegetation.

The Prepuna

The Prepuna (Czajka and Vervoorst, 1956; Cabrera, 1971) extends along the dry mountain slopes of northwestern Argentina from the state of Jujuy to La Rioja, approximately between 2,000 and 3,400 m. It is characterized by a dry and warm climate with summer rains; it is warmer than the Puna, colder than the Monte; and it is a special formation strongly influenced by the exposure of the mountains in the region.

The vegetation is mainly formed by xerophytic shrubs and cacti. Among the shrubs, the most abundant are *Gochnatia glutinosa* (Compositae), *Cassia crassiramea* (Leguminosae), *Aphyllocladus spartioides, Caesalpinia trichocarpa* (Leguminosae), *Proustia cuneifolia* (Compositae), *Chuquiraga erinacea* (Compositae), *Zuccagnia punctata* (Leguminosae), *Adesmia inflexa* (Leguminosae), and *Psila boliviensis* (Compositae). The most conspicuous member of the Cactaceae is the cardon, *Trichocereus pasacana*; there are also present *T. poco* and species of *Opuntia, Cylindropuntia, Tephrocactus, Parodia*, and *Lobivia*. Among the grasses are *Digitaria californica, Stipa leptostachya, Monroa argentina*, and *Agrostis nana*.

The Pacific Coastal Desert

Along the Peruvian and Chilean coast from lat. 5° S to approximately lat. 30° S, we find the region denominated "La Costa" in Peru (Weberbauer, 1945; Ferreyra, 1960) and "Northern Desert," "Coastal Desert," or "Atacama Desert" in Chile (Johnston, 1929; Reiche, 1934; Goodspeed, 1945). This very dry region is under the influence of the combined rain shadow of the high cordillera to the east and the cold Humboldt Current and the coastal upwelling along the Peruvian coast. Although physically continuous, the vegetation is not uniform, as a result of the combination of temperature and rainfall variations in such an extended territory. Temperature decreases from north to south as can be expected, going from a yearly mean to close to 25°C in northern Peru (Ferreyra, 1960) to a low of 15°C at its southern border. Rainfall is very irregular and very meager. Although some localities in Peru (Zorritos, Lomas de Lachay; cf. Ferreyra, 1960) have averages of 200 mm, the average yearly rainfall is below 50 mm in most places. This has created an extreme xerophytic vegetation often with special adaptations to make use of the coastal fog.

Behind the coastal area are a number of dry valleys, some in Peru but mostly in northern Chile, with the same kind of extreme dry conditions as the coastal area.

The flora is characterized by plants with extreme xerophytic adaptations, especially succulents, such as *Cereus spinibaris* and *C. coquimbanus*, various species of *Echinocactus*, and *Euphorbia lactifolia*. The most interesting associations occur in the so-called *lomas*, or low hills (less than 1,500 m), along the coast that intercept the coastal fog and provide very localized conditions favorable for some plant growth. Almost each of these formations from the Ecuadorian border to central Chile constitutes a unique community. Over 40 percent of the plants in the Peruvian coastal community are annuals (Ferreyra, 1960), although annuals apparently are less common in Chile (Johnston, 1929); of the perennials, a large number are root perennials or succulents. Only about 5 percent are shrubs or trees in the northern sites (Ferreyra, 1960), while shrubs and semishrubs constitute a higher proportion in the Chilean region. From the Chilean region should be mentioned *Oxalis gigantea* (Oxalidaceae), *Heliotropium philippianum* (Boraginaceae), *Salvia gilliesii* (Labiatae), and

Proustia tipia (Compositae) among the shrubs; species of *Poa*, *Eragrostis*, *Elymus*, *Stipa*, and *Nasella* among the grasses; and *Alstroemeria violacea* (Amaryllidaceae), a conspicuous and relatively common root perennial. In southern Peru *Nolana inflata*, *N. spathulata* (Nolanaceae), and other species of this widespread genus; *Tropaeolum majus* (Tropaeolaceae), *Loasa urens* (Loasaceae), and *Arcythophyllum thymifolium* (Rubiaceae); in the *lomas* of central Peru the *amancay*, *Hymenocallis amancaes* (Amaryllidaceae), *Alstroemeria recumbens* (Amaryllidaceae), *Peperomia atocongona* (Piperaceae), *Vicia lomensis* (Leguminosae), *Carica candicans* (Caricaceae), *Lobelia decurrens* (Lobeliaceae), *Drymaria weberbaueri* (Caryophyllaceae), *Capparis prisca* (Capparidaceae), *Caesalpinia tinctoria* (Leguminosae), *Pitcairnia lopezii* (Bromeliaceae), and *Haageocereus lachayensis* and *Armatocereus* sp. (Cactaceae). Finally, in the north we find *Tillandsia recurvata*, *Fourcroya occidentalis*, *Apralanthera ferreyra*, *Solanum multinterruptum*, and so on.

Of great phytogeographic interest is the existence of a less-xerophytic element in the very northern extreme of the Pacific Coastal Desert, from Trujillo to the border with Ecuador (Ferreyra, 1960), known as *algarrobal*. Principal elements of this vegetation are two species of *Prosopis*, *P. limensis* and *P. chilensis*; others are *Cercidium praecox*, *Caesalpinia paipai*, *Acacia huarango*, *Bursera graveolens* (Burseraceae), *Celtis iguanea* (Ulmaceae), *Bougainvillea peruviana* (Nyctaginaceae), *Cordia rotundifolia* (Boraginaceae), and *Grabowskia boerhaviifolia* (Solanaceae).

Geological History

The present desert and subdesert regions of temperate South America result from the existence of belts of high atmospheric pressure around lat. 30° S, high mountain chains that impede the transport of moisture from the oceans to the continents, and cold water currents along the coast, which by cooling and drying the air that flows over them act like the high mountains.

The Pacific Coastal Desert of Chile and Peru is principally the result of the effect of the cold Humboldt Current that flows from south to north; the Patagonian Steppe is produced by the Cordillera de los Andes that traps the moisture in the prevailing westerly winds; while the Monte and the Puna result from a combination of the cordilleran rain shadow in the west and the Sierras Pampeanas in the east and the existence of the belt of high pressure.

The high-pressure belt of mid-latitudes is a result of the global flow of air (Flohn, 1969) and most likely has existed with little modification throughout the Mesozoic and Cenozoic (however, for a different view, see Schwarzenbach, 1968, and Volkheimer, 1971). The mountain chains and the cold currents, on the other hand, are relatively recent phenomena. The latter's low temperature is largely the result of Antarctic ice. But aridity results from the interaction of temperature and humidity. In effect, when ambient temperatures are high, a greater percentage of the incident rainfall is lost as evaporation and, in addition, plants will transpire more water. Consequently, in order to reconstruct the history of the desert and semidesert regions of South America, we also have to have an idea of the temperature and pluvial regimes of the past.

In this presentation I will use two types of evidence: (1) the purely geological evidence regarding continental drift, times of uplifting of mountain chains, marine transgressions, and existence of paleosoils and pedemonts; and (2) paleontological evidence regarding the ecological types and phylogenetical stock of the organisms that inhabited the area in the past. With this evidence I will try to reconstruct the most likely climate for temperate South America since the Cretaceous and deduce the kind of vegetation that must have existed.

Cretaceous

This account will start from the Cretaceous because it is the oldest period from which we have fossil records of angiosperms, which today constitute more than 90 percent of the vascular flora of the regions under consideration. At the beginning of the Cretaceous, South America and Africa were probably still connected (Dietz and Holden, 1970), since the rift that created the South Atlantic and separated the two continents apparently had its origin during the Lower Cretaceous. The position of South America at this time was slightly south (approximately lat. 5°–10° S) of its present position and with its southern extremity tilted eastward. There were no significant mountain chains at that time.

Northern and western South America are characterized in the Cretaceous by extensive marine transgressions in Colombia, Venezuela, Ecuador, and Peru (Harrington, 1962). In Chile, during the middle Cretaceous, orogeny and uplift of the Chilean Andes began (Kummel, 1961). This general zone of uplift, which was accompanied by active volcanism and which extended to central Peru, marks the beginning of the formation of the Andean cordillera, a phenomenon that will have its maximum expression during the upper Pliocene and Pleistocene and that is not over yet.

Although the first records of angiosperms date from the Cretaceous (Maestrichtian), the known fossil floras from the Cretaceous of South America are formed predominantly by Pteridophytes, Bennettitales, and Conifers (Menéndez, 1969). Likewise, the fossil faunas are formed by dinosaurs and other reptilian groups. Toward the end of the Cretaceous (or beginning of Paleocene) appear the first mammals (Patterson and Pascual, 1972).

Climatologically, the record points to a much warmer and possibly wetter climate than today, although there is evidence of some aridity, particularly in the Lower Cretaceous.

All in all, the Cretaceous period offers little conclusive evidence of extensive dry conditions in South America. Nevertheless, during the Lower Cretaceous before the formation of an extensive South Atlantic Ocean, conditions in the central portion of the combined continent must have been drier than today. In effect, the high rainfall in the present Amazonian region is the result of the condensation of moisture from rising tropical air that is cooling adiabatically. This air is brought in by the trade winds and acquires its moisture over the North and South Atlantic. Before the breakup of Pangea, trade winds must have been considerably drier on the western edge of the continent after blowing over several thousand miles of hot land. It is interesting that some characteristic genera of semidesert regions, such as *Prosopis* and *Acacia*, are represented in both eastern Africa and South America. This disjunct distribution can be interpreted by assuming Cretaceous origin for these genera, with a more or less continuous

Cretaceous distribution that was disrupted when the continents separated (Thorne, 1973). This is in accordance with their presumed primitive position within the Leguminosae (L. I. Nevling, 1970, personal communication). There is some geomorphological evidence also for at least local aridity in the deposits of the Lower Cretaceous of Córdoba and San Luis in Argentina, which are of a "typical desert phase" according to Gordillo and Lencinas (1972).

Cenozoic

Paleocene. The marine intrusions of northern South America still persisted at the beginning of the Paleocene but had become much less extensive (Haffer, 1970). The Venezuelan Andes and part of the Caribbean range of Venezuela began to rise above sea level (Liddle, 1946; Harrington, 1962). In eastern Colombia, Ecuador, and Peru continental deposits were laid down to the east of the rising mountains, which at this stage were still rather low. The sea retreated from southern Chile, but there was a marine transgression in central eastern Patagonia.

At the beginning of the Paleocene the South American flora acquired a character of its own, very distinct from contemporaneous European floras, although there are resemblances to the African flora (Van der Hammen, 1966). The first record of Bombacaceae is from this period (Van der Hammen, 1966).

There are remains of crocodiles from the Paleocene of Chubut in Argentina, indicating a probable mean temperature of 10°C or higher for the coldest month (Volkheimer, 1971), some fifteen to twenty degrees warmer than today. The early Tertiary mammalian fossil faunae consisted of marsupials, edentates of the suborder Xenarthra, and a variety of ungulates (Patterson and Pascual, 1972). These forms appear to have lived in a forested environment, confirming the paleobotanical evidence (Menéndez, 1969, 1972; Petriella, 1972).

The climate of South America during the Paleocene was clearly warmer and more humid than today. With the South Atlantic now fairly large and with no very great mountain range in existence, probably no extensive dry-land floras could have existed.

Eocene. During the Eocene the general features of the northern Andes were little changed from the preceding Paleocene. The northern extremity of the eastern cordillera began to be uplifted. In western Colombia and Ecuador the Bolívar geosyncline was opened (Schuchert, 1935; Harrington, 1962). Thick continental beds were deposited in eastern Colombia-Peru, mainly derived from the erosion of the rising mountains to the east. In the south the slow rising of the cordillera continued. There was an extensive marine intrusion in eastern Patagonia.

The flora was predominantly subtropical (Romero, 1973). It was during the Eocene that the tropical elements ranged farthest south, which can be seen very well in the fossil flora of Río Turbio in Argentine Patagonia. Here the lowermost beds containing *Nothofagus* fossils are replaced by a rich flora of tropical elements with species of *Myrica*, *Persea*, *Psidium*, and others, which is then again replaced in still higher beds by a *Nothofagus* flora of more mesic character (Hünicken, 1966; Menéndez, 1972).

However, in the Eocene we also find the first evidence of elements belonging to a more open, drier vegetation, particularly grasses (Menéndez, 1972; Van der Hammen, 1966).

The Eocene was also a time of radiation of several mammalian phyletic lines, particularly marsupials, xenarthrans, ungulates, and notoungulates (Patterson and Pascual, 1972). Of particular interest for our purpose is the appearance of several groups of large native herbivores (Patterson and Pascual, 1972). More interesting still is "the precocity shown by certain ungulates in the acquisition of high-crowned, or hyposodont, and rootless, or hypselodont, teeth" (Patterson and Pascual, 1972). By the lower Oligocene such teeth had been acquired by no fewer than six groups of ungulates. Such animals must have thrived in the evolving pampas areas. True pampas are probably younger, but by the Eocene it seems reasonable to propose the existence of open savanna woodlands, somewhat like the llanos of Venezuela today.

The climate appears to have been fairly wet and warm until a peak was reached in middle Eocene, after which time a very gradual drying and cooling seems to have occurred.

Oligocene. The geological history of South America during the Oligocene followed the events of the earlier periods. There were further uplifts of the Caribbean and Venezuelan mountains and also the Cordillera Principal of Peru. In Patagonia the cordillera was uplifted and the coastal cordillera also began to rise. At the same time, erosion of these mountains was taking place with deposition to the east of them.

In Patagonia elements of the Eocene flora retreated northward and the temperate elements of the *Nothofagus* flora advanced. In northern South America all the evidence points to a continuation of a tropical forest landscape, although with a great deal of phyletic evolution (Van der Hammen, 1966).

The paleontological record of mammals shows the continuing radiation and gradual evolution of the stock of ancient inhabitants of South America. The Oligocene also records the appearance of caviomorph rodents and platyrrhine primates, which probably arrived from North America via a sweepstakes route (Simpson, 1950; Patterson and Pascual, 1972), although an African origin has also been proposed (Hoffstetter, 1972).

Miocene. During Miocene times a number of important geological events took place. In the north the eastern cordillera of Colombia, which had been rising slowly since the beginning of the Tertiary, suffered its first strong uplift (Harrington, 1962). The large deposition of continental deposits in eastern Colombia and Peru continued, and by the end of the period the present altiplano of Peru and Bolivia had been eroded almost to sea level (Kummel, 1961). In the southern part of the continent one sees volcanic activity in Chubut and Santa Cruz as well as continued uplifting of the cordillera. By the end of the Miocene we begin to see the rise of the eastern and central cordilleras of Bolivia and the Puna and Pampean ranges of northern Argentina (Harrington, 1962).

During the Miocene the southern *Nothofagus* forest reached an extension similar to that of today. By the end of this time the pampa, large grassy extensions in central Argentina, became quite widespread (Patterson and Pascual, 1972). We also see the appearance and radiation of Compositae, a typical element in nonforested areas today (Van der Hammen, 1966). Among the fauna no major changes took place.

The climate continued to deteriorate from its peak of wet-warm in the middle Eocene. It still was more humid than today, as the presence of thick paleosoils in Patagonia seem to indicate (Volkheimer, 1971).

Nevertheless, the southern part of the continent, other than locally, was no longer occupied by forest but most certainly by either grassland or a parkland. The reduced rainfall, together with the ever-increasing rain-shadow effect of the rising Andes, must have led to long dry seasons in the middle latitudes. Indirect evidence from the evolutionary history of some bird and frog groups appears to indicate that the *Nothofagus* forest was not surrounded by forest vegetation at this time (Hecht, 1963; Vuilleumier, 1967). It is also very likely that semidesert regions existed in intermountain valleys and in the lee of the rising mountains in the western part of the continent from Patagonia northward.

Pliocene. From the Pliocene we have the first unmistakable evidence for the existence of more or less extensive areas of semidesert. Geologically it was a very active period. In the north we see the elevation of the Bolívar geosyncline and the development of the Colombian Andes in their present form, leading to the connection of South and North America toward the end of the period (Haffer, 1970). In Peru we see the rising of the cordillera and the bodily uplift of the altiplano to its present level, followed by some rifting. In Chile and Argentina we see the beginning of the final rise of the Cordillera Central as well as the uplift of the Sierras Pampeanas and the precordillera. All this increased orogenic activity was accompanied by extensive erosion and the deposition of continental sediments to the east in the Amazonian and Paraná-Paraguay basins (Harrington, 1962).

The lowland flora of northern South America, particularly that of the Amazonas and Orinoco basins, was not too different from today's flora in physiognomy or probably in floristic composition. However, because of the rise of the cordillera we find in the Pliocene the first indications of the existence of a high mountain flora (Van der Hammen, 1966) as well as the first clear indication of the existence of desert vegetation (Simpson Vuilleumier, 1967; Van der Hammen, 1966).

With the disappearance of the Bolívar geosyncline in late Pliocene, South America ceased to be an island and became connected to North America. This had a very marked influence on the fauna of the continent (Simpson, 1950; Patterson and Pascual, 1972). In effect, extensive faunistic interchanges took place during the Pliocene and Pleistocene between the two continents.

By the end of the Pliocene the landscape of South America was essentially identical to its present form. The rise of the Peruvian and Bolivian areas that we know as the Puna had taken place creating the dry highlands; the uplift of the Cordillera Central of Chile and the Sierras Pampeanas of Argentina had produced the rain shadows that make the area between them the dry land it is; and, finally, the rise of the southern cordillera of Chile must have produced dry, steppelike conditions in Patagonia. Geomorphological evidence shows this to be true (Simpson Vuilleumier, 1967; Volkheimer, 1971). The coastal region of Chile and Peru was probably more humid than today, since the cold Humboldt Current probably did not exist yet in its present form (Raven, 1971, 1973). However, although the stage is set, the actors are not quite ready. In effect, the Pleistocene, although very short in duration compared to the Tertiary events just described, had profound effects on species formation and distribution by drastically affecting the climate. Furthermore, because of its recency we also have a much better geological and paleontological record and therefore knowledge of the events of the Pleistocene.

Pleistocene

The deterioration of the Cenozoic climate culminated in the Pleistocene, when temperatures in the higher latitudes were lowered sufficiently to allow the accumulation and spread of immense ice sheets in the northern continents and on the highlands of the southern continents. Four major glacial periods are usually recognized in the Northern Hemisphere (Europe and North America), with three milder interglacial periods between them, and a fourth starting about 10,000 B.P. (Holocene), in which we are presently living. It is generally agreed (Charlesworth, 1957; Wright and Frey, 1965; Frenzel, 1968) that the Pleistocene has been a time of great variations in climate, both in temperature and in humidity, associated with rather significant changes in sea level (Emiliani, 1966). In general, glacial maxima correspond to colder and wetter climates than exist today; interglacials to warmer and often drier periods. But the march of events was more complex, and the temperature and humidity changes were not necessarily correlated (Charlesworth, 1957). Neither the exact series of events nor their ultimate causes are entirely clear.

Simpson Vuilleumier (1971), Van der Hammen and González (1960), and Van der Hammen (1961, 1966) have reviewed the Pleistocene events in South America. In northern South America (Venezuela, Colombia, and Ecuador) one to three glaciations took place, corresponding to the last three events in the Northern Hemisphere (Würm, Riss, and Mindel). In Peru, Bolivia, northern Chile, and Argentina there were at least three, in some areas possibly four. In Patagonia there were three to four glaciation events (table 2-1). All these glaciations, with the possible exception of Patagonia (Auer, 1960; Czajka, 1966), were the result of mountain glaciers.

The alternation of cold, wet periods with warm-dry and warm-wet periods had drastic effects on the biota. During glacial periods snow lines were lowered with an expansion of the areas suitable for a high mountain vegetation (Van der Hammen, 1966; Simpson Vuilleumier, 1971). At the same time glaciers moving along valleys created barriers to gene flow in some cases. During interglacials the snow line moved up again, and the areas occupied by high mountain vegetation no doubt were interrupted by low-lying valleys, which were occupied by more mesic-type plants. On the other hand, particularly at the beginning of interglacials, large mountain lakes were produced, and later on, with the rise of sea level, marine intrusions appeared. These events also broke up the ranges of species and created barriers to gene flow. To these happenings have to be added the effects of varying patterns of aridity and humidity. Let us then briefly review the events and their possible effects on the semidesert areas of temperate South America.

Patagonia. Glacial phenomena are best known from Patagonia (Caldenius, 1932; Feruglio, 1949; Frenguelli, 1957; Auer, 1960; Czajka, 1966; Flint and Fidalgo, 1968). Three or four glacial events are recorded. Along the cordillera the *Nothofagus* forest retreated north. The ice in its maximum extent covered probably most of Tierra del Fuego, all the area west of the cordillera, and some 100 km east of the mountains. Furthermore, during the glacial maxima, as a result of the lowering of the sea level, the Patagonian coastline was situated almost 300 km east of its present position. The climate was definitely colder and more humid. Studies by Auer (1958, 1960) indicate, however, that the *Nothofagus* forest did not expand eastward to any con-

Table 2-1. *Summary of Glacial Events in South America*

Localities	Glaciations (no.)	Age of Glaciations Relative to Europe	Present Snow Line (m)
Venezuela Mérida, Perija	1 or 2	Würm or Riss & Würm	4,800–4,900
Colombia Santa Marta and Cordillera C.	Variable, 1 to 3	Mindel to Würm	4,200–4,500
Peru All high Andean peaks	3	Mindel to Würm	5,800(W); 5,000(E)
Bolivia All NE ranges; high peaks in SE; few peaks in SW	3 or 4	Günz or Mindel to Würm	5,900(W); ca. 5,300(E)
Argentina and Chile Peaks between lat. 30° and 42°S; all land to the west of main Andean chain; to the east only to the base of the cordillera	3 or 4	Günz or Mindel to Würm	Variable: above 5,900 m in north to 800 m in south
Paraguay, Brazil, and Argentina Paraná basin		no glaciations	
Brazil Mt. Itatiaia	1 or 2	Würm or Riss & Würm	none
Brazil Amazonas basin		no glaciations	

Source: Modified from Simpson Vuilleumier, 1971.

Glacial Snow Line (m)	Glacial Climate	Interglacial Climate
2,700–3,300		
4,500(W); 4,200(E)	wet, temp. 4° to 11° lower than present	dry, temp. 2° to 3° higher than present
4,500(W); 4,200(E)	wet, temp. 7° lower than present	
5,000–5,300(W); 4,600–5,000(E)	wet, temp. 6° lower than present	
500 m at Santiago, Chile; sea level south of lat. 42°	wet	more genial than present
	cool, dry	humid, warm
2,300		
	cool, dry	humid, warm

siderable extent. It must be remembered that, even though the climate was more humid, the prevailing winds still would have been westerlies and they still would have discharged most of their humidity when they collided with the cordillera as is the case today. The drastically lowered snow line and the cold-dry conditions of Patagonia, on the other hand, must have had the effect of allowing the expansion of the high mountain flora that began to evolve as a result of the uplift of the cordillera in the Pliocene and earlier.

Monte. The essential semidesert nature of the Monte region was probably not affected by the events of the Pleistocene, but the extent of the area must have fluctuated considerably during this time. In effect, during glacial maxima not only did some regions become covered with ice, such as the valley of Santa María in Catamarca, but they also became colder. On the other hand, during interglacials there is evidence for a moister climatic regime, as the existence of fossil woodlands of *Prosopis* and *Aspidosperma* indicates (Groeber, 1936; Castellanos, 1956). Also, the present patterns of distribution of many mesophytic (but not wet-tropical) species or pairs of species, with populations in southern Brazil and the eastern Andes, could probably only have been established during a wetter period (Smith, 1962; Simpson Vuilleumier, 1971). On the other hand, geomorphological evidence from the loess strata of the Paraná-Paraguay basin (Padula, 1972) shows that there were at least two periods when the basin was a cool, dry steppe. During these periods the semidesert Monte vegetation must have expanded northward and to the east of its present range.

Puna. It has already been noted that during glacial maxima the snow line was lowered and the area open for colonization by the high mountain elements was considerably extended. Nowhere did that become more significant than in the Puna area (Simpson Vuilleumier, 1971). During glacial periods a number of extensive glaciers were formed in the mountains surrounding it, particularly the Cordillera Real near La Paz (Ahlfeld and Branisa, 1960). Numerous and extensive glacial lakes were also formed (Steinmann, 1930; Ahlfeld and Branisa, 1960; Simpson Vuilleumier, 1971). However, the basic nature of the Puna vegetation was probably not affected by these events. They must, however, have produced extensive shifts in ranges and isolation of populations, events that must have increased the rate of evolution and speciation.

Pacific Coastal Desert. The Pacific Coastal Desert is the result of the double rain shadow produced by the Andes to the east and the cold Humboldt Current to the west. The Andes did not reach their present size until the end of the Pliocene or later. The cold Humboldt Current did not become the barrier it is until its waters cooled considerably as a result of being fed by melt waters of Antarctic ice. The coastal cordillera, however, was higher in the Pleistocene than it is today (Cecioni, 1970). It is not possible to state categorically when the conditions that account for the Pacific Coastal Desert developed, but it was almost certainly not before the first interglacial. Consequently, it is safe to say that the Pacific Coastal Desert is a Pleistocene phenomenon, as is the area of Mediterranean climate farther south (Axelrod, 1973; Raven, 1973).

During the Pleistocene the snow line in the cordillera was considerably depressed and may have been as low as 1,300 m in some places (Simpson Vuilleumier, 1971). Estimates of temperature depressions are in the order of 7°C (Ahlfeld and Branisa, 1960). Although the ice did not reach the coast, the lowered temperature probably resulted in a much lowered timber line and expansion of Andean elements into the Pacific Coastal Desert. There is also evidence for dry and humid cycles during interglacial periods (Simpson Vuilleumier, 1971). The cold glacial followed by the dry interglacial periods probably decimated the tropical and subtropical elements that occupied the area in the Tertiary and allowed the invasion and adaptive radiation of cold- and dry-adapted Andean elements.

Holocene

We finally must consider the events of the last twelve thousand years, which set the stage for today's flora and vegetation. Evidence from Colombia, Brazil, Guyana, and Panama (Van der Hammen, 1966; Wijmstra and Van der Hammen, 1966; Bartlett and Barghoorn, 1973) indicates that the period started with a wet-warm period that lasted for two to four thousand years, followed by a period of colder and drier weather that reached approximately to 4,000 B.P. when the forest

retreated, after which present conditions gradually became established. The wet-humid periods were times of expansion of the tropical
vegetation, while the dry period was one of retreat and expansion of
savannalike vegetation which appears to have occupied extensive
areas of what is today the Amazonian basin (Van der Hammen, 1966;
Haffer, 1969; Vanzolini and Williams, 1970; Simpson Vuilleumier,
1971). Unfortunately, no such detailed observations exist for the
temperate regions of South America, but it is likely that the same alter-
nations of wet, dry, and wet took place there, too.

The Floristic Affinities

Cabrera (1971) divides the vegetation of the earth into seven major
regions, two of which, the *Neotropical* and *Antarctic* regions, include
the vegetation of South America. The latter region comprises in South
America only the area of the *Nothofagus* forest along both sides of the
Andes from approximately lat. 35° S to Antarctica and the subantarctic
islands (fig. 2-1). The Neotropical region, which occupies the rest of
South America, is divided further into three dominions comprising,
broadly speaking, the tropical flora (Amazonian Dominion), the subtropical vegetation (Chaco Dominion), and the vegetation of the
Andes (Andean-Patagonian Dominion). The Chaco Dominion is
further subdivided into seven phytogeographical provinces. Two of
these are semidesert regions: the Monte province and the Prepuna
province. The remaining five provinces of the Chaco Dominion are
the Matorral or central Chilean province, the Chaco province, the
Argentine Espinal (not to be confused with the Chilean Matorral), the
region of the Pampa, and the region of the Caatinga in northeastern
Brazil. With exception of the Matorral and the Caatinga, the other
provinces of the Chaco Dominion are contiguous and reflect a different set of temperature, rainfall, and soil conditions in each case.
The other dominion of the Neotropical flora that concerns us here is
the Andean-Patagonian one, with three provinces: Patagonia, the
Puna, and the vegetation of the high mountains. We see then that, of
the five subdesert temperate provinces, two have a flora that is subtropical in origin and three a flora that is related to the high mountain

vegetation. We will now briefly discuss the floristic affinities of each of these regions.

The Monte

The vegetation, flora, and floristic affinities of the Monte are the best known of all temperate semidesert regions (Vervoorst, 1945, 1973; Czajka and Vervoorst, 1956; Morello, 1958; Sarmiento, 1972; Solbrig, 1972, 1973). There is unanimous agreement that the flora of the Monte is related to that of the Chaco province (Cabrera, 1953, 1971; Sarmiento, 1972; Vervoorst, 1973).

Sarmiento (1972) and Vervoorst (1973) have made statistical comparisons between the Chaco and the Monte. Sarmiento, using a number of indices, shows that the Monte scrub is most closely related, both floristically and ecologically, to the contiguous dry Chaco woodland. Vervoorst, using a slightly different approach, shows that certain Monte communities, particularly on mountain slopes, have a greater number of Chaco species than other more xerophytic communities, particularly the *Larrea* flats and the vegetation of the sand dunes. Altogether, better than 60 percent of the species and more than 80 percent of the genera of the Monte are also found in the Chaco.

The most important element in the Monte vegetation is the genus *Larrea* with four species. Three of these—*L. divaricata*, *L. nitida*, and *L. cuneifolia*—constitute the dominant element over most of the surface of the Monte, either singly or in association (Barbour and Díaz, 1972; Hunziker et al., 1973). The fourth species, *L. ameghinoi*, a low-creeping shrub, is found in depressions on the southern border of the Monte and over extensive areas of northern Patagonia. Of the three remaining species, *L. cuneifolia* is found in Chile in the area between the Matorral and the beginning of the Pacific Coastal Desert, known locally as Espinal (not to be confused with the Argentine Espinal). *Larrea divaricata* has the widest distribution of the species in the genus. In Argentina it is found throughout the Monte as well as in the dry parts of the Argentine Espinal and Chaco up to the 600-mm isohyet (Morello, 1971, personal communication). However, there is some question whether the present distribution of *L. divaricata* in the Chaco is natural or the result of the destruction of the natural vege-

tation by man since *L. divaricata* is known to be invasive. This species
is also found in Chile in the central provinces and in two isolated localities in Bolivia and Peru: the valley of Chuquibamba in Peru and the
region of Tarija in Bolivia (Morello, 1958; Hunziker et al., 1973).
Finally, *L. divaricata* is found in the semidesert regions of North America from Mexico to California (Yang, 1970; Hunziker et al., 1973).

The second most important genus in the Monte is *Prosopis*. Of the
species of *Prosopis* found there, two of the most important ones (*P.
alba*, *P. nigra*) are characteristic species in the Chaco and Argentine
Espinal where they are widespread and abundant. A third very characteristic species of *Prosopis*, *P. chilensis*, is found in central and
northern Chile, in the Matorral where it is fairly common and in some
interior localities of the Pacific Coastal Desert, as well as in northern
Peru. The records of *P. chilensis* from farther north in Ecuador and
Colombia, and even from Mexico, correspond to the closely related
species *P. juliflora*, considered at one time conspecific with *P. chilensis* (Burkart, 1940, 1952). *Prosopis alpataco* is found in the Monte and
in Patagonia. Most other species of the genus have more limited distributions.

Another conspicuous element in the Monte is *Cercidium*. The
genus is distributed from the semidesert regions in the United States
where it is an important element of the flora, south along the Cordillera
de los Andes, with a rather large distributional gap in the tropical region from Mexico to Ecuador. *Cercidium* is found in dry valleys of the
Pacific Coastal Desert and in the cordillera in Peru and Chile. In Argentina it is found, in addition to the Monte, in the western edge of the
Chaco, in the Prepuna, and also in the Puna (Johnston, 1924).

Bulnesia is represented in the Monte by two species, *B. retama*
and *B. schickendanzii*. The first of these species is found also in the
Pacific Coastal Desert in the region of Ica and Nazca, *B. schickendanzii*, however, is a characteristic element of the Prepuna province.
Other interesting distributions among characteristic Monte species
are the presence of *Bougainvillea spinosa* in the department of
Moquegua in Peru (where it grows with *Cercidium praecox*). The
highly specialized *Monttea aphylla* is endemic to the Monte, but a very
closely related species, *Monttea chilensis*, is found in northern Chile.
Geoffroea decorticans, the *chañar*, which is an important element

both in the Chaco and in the Monte, is also found in northern Chile where it is common. These are but a few of the more important examples of Monte species that range into other semidesert phytogeographical provinces, particularly the Pacific Coastal Desert.

In summary, the Monte has its primary floristic connection with the Chaco but also has species belonging to an Andean stock. In addition, a number of important Monte elements are found in isolated dry pockets in southern Bolivia (Tarija), northern Chile, and coastal Peru and are hard to classify.

The Prepuna

There are no precise studies on the flora or the floristic affinities of the Prepuna. However, a look at the common species indicates a clear affinity with the Chaco and the Monte, such as *Zuccagnia punctata* (Monte), *Bulnesia schickendanzii* (Monte), *Bougainvillea spinosa* (Monte), *Trichocereus tertscheckii* (Monte), and *Cercidium praecox* (Monte and Chaco). Other elements are clearly Puna elements: *Psila boliviensis*, *Junellia juniperina*, and *Stipa leptostachya*. Although the Prepuna province has a physiognomy and floral mixture of its own, it undoubtedly has a certain ecotone nature, and its limits and its individuality are most probably Holocene events.

The Puna and Patagonia

Although the floristic affinities of the Puna and Patagonia have not been studied in as much detail as those of the Monte, they do not present any special problem. The flora of both regions is clearly part of the Andean flora. This important South American floristic element is relatively new (since it cannot be older than the Andes). This is further shown by the paucity of endemic families (only two small families, Nolanaceae [also found in the Galápagos Islands] and Malesherbiaceae, are endemic to the Andean Dominion) and by the large number of taxa belonging to such families as Compositae, Gramineae, Verbenaceae, Solanaceae, and Cruciferae, considered usually to be relatively specialized and geologically recent. The Leguminosae, represented in the Chaco Dominion mostly by Mimosoideae (among them some primitive genera), are chiefly represented in the Andean Dominion by more advanced and specialized genera of the Papilionoideae.

The Patagonian Steppe is characterized by a very large number of endemic genera, but particularly of endemic species (over 50%, cf. Cabrera, 1947). Of the species whose range extends beyond Patagonia, the great majority grow in the cordillera, a few extend into the *Nothofagus* forest, and a very small number are shared with the Monte. This is surprising in view of some similarities in soil and water stress between the two regions and also in view of the lack of any obvious physical barrier between the two phytogeographical provinces.

The Pacific Coastal Desert

The flora of the Pacific Coastal Desert is the least known. The relative lack of communications in this region, the almost uninhabited nature of large parts of the territory, and the harshness of the climate and the physical habitat have made exploration very difficult. Furthermore, a large number of species in this region are ephemerals, growing and blooming only in rainy years. Our knowledge is based largely on the works of Weberbauer and Ferreyra in Peru and those of Philippi, Johnston, and Reiche in Chile.

One of the characteristics of the region is the large number of endemic taxa. The only two endemic families of the Andean Dominion, the Malesherbiaceae and the Nolanaceae, are found here; many of the genera and most of the species are also endemic.

The majority of the species and genera are clearly related to the Andean flora. The common families are Compositae, Umbelliferae, Cruciferae, Caryophyllaceae, Gramineae, and Boraginaceae, all families that are considered advanced and geologically recent. In this it is similar to Patagonia. However, the region does not share many taxa with Patagonia, indicating an independent history from Andean ancestral stock, as is to be expected from its geographical position.

On the other hand, contrary to Patagonia, the Pacific Coastal Desert has elements that are clearly from the Chaco Dominion. Among them are *Geoffroea decorticans*, *Prosopis chilensis*, *Acacia caven*, *Zuccagnia punctata*, and pairs of vicarious species in *Monttea* (*M. aphylla*, *M. chilensis*), *Bulnesia* (*B. retama*, *B. chilensis*), *Goch-*

natia, and *Proustia*. In addition there are isolated populations of *Bulnesia retama*, *Bougainvillea spinosa*, and *Larrea divaricata* in Peru. Because the Monte and the Pacific Coastal Desert are separated today by the great expanse of the Cordillera de los Andes that reaches to over 5,000 m and by a minimum distance of 200 km, these isolated populations of Monte and Chaco plants are very significant.

Discussion and Conclusions

In the preceding pages a brief description of the desert and semi-desert regions of temperate South America was presented, as well as a short history of the known major geological and biological events of the Tertiary and Quaternary and the present-day floristic affinities of the regions under consideration. An attempt will now be made to relate these facts into a coherent theory from which some verifiable predictions can be made.

The paleobotanical evidence shows that the Neotropical flora and the Antarctic flora were distinct entities already in Cretaceous times (Menéndez, 1972) and that they have maintained that distinctness throughout the Tertiary and Quaternary in spite of changes in their ranges (mainly an expansion of the Antarctic flora). The record further indicates that the Antarctic flora in South America was always a geographical and floristic unit, being restricted in its range to the cold, humid slopes of the southern Andes. The origin of this flora is a separate problem (Pantin, 1960; Darlington, 1965) and will not be considered here. Some specialized elements of this flora expanded their range at the time of the lifting of the Andes (*Drimys*, *Lagenophora*, etc.), but the contribution of the Antarctic flora to the desert and semidesert regions is negligible. The discussion will be concerned, therefore, exclusively with the Neotropical flora from here on.

The data suggest that at the Cretaceous-Tertiary boundary (between Maestrichtian and Paleocene) the Neotropical angiosperm flora covered all of South America with the exception of the very southern tip. The evidence for this assertion is that the known fossil floras of that time coming from southern Patagonia (Menéndez, 1972) indicate the existence then of a tropical, rain-forest-type flora in a region that today supports xerophytic, cold-adapted scrub and cushion-plant vegetation. The reasoning is that if at that time it was hot and humid enough in the southernmost part of the continent for a rain forest, undoubtedly such conditions would be more prevalent farther north. Such reasoning, although largely correct, does not take into account all the factors.

If we accept that the global flow of air and the pattern of insolation of the earth were essentially the same throughout the time under consideration (see "Theory"), it is reasonable to assume that a gradient of increasing temperature from the poles to the equator was in existence. But it is not necessarily true that a similar gradient of humidity existed. In effect, on a perfect globe (one where the specific heat of water and land is not a factor) the equatorial region and the middle high latitudes (around 40°–60°) would be zones of high rainfall while the middle latitudes (25°–30°) and the polar regions would be regions of low rainfall. This is the consequence of the global movements of air (rising at the poles and middle high latitudes and consequently cooling adiabatically and discharging their humidity, falling in the middle latitudes and the poles and consequently heating and absorbing humidity). But the earth is not a perfect globe, and, consequently, the effects of distribution of land masses and oceans have to be taken into account. When air flows over water, it picks up humidity; when it flows over land, it tends to discharge humidity; when it encounters mountains, it rises, cools, and discharges humidity; behind a mountain it falls and heats and absorbs humidity (which is the reason why Patagonia is a semidesert today).

As far as can be ascertained, at the beginning of the Tertiary there were no large mountain chains in South America. Therefore the expected air flow probably was closer to the ideal, that is, humid in the tropics and in the middle low latitudes, relatively dry in mid-latitudes. I would like to propose, therefore, that at the beginning of the Tertiary South America was not covered by a blanket of rain forest, but that at middle latitudes, particularly in the western part of the continent, there existed a tropical (since the temperature was high) flora adapted to a seasonally dry climate. This was not a semidesert flora but most likely a deciduous or semideciduous forest with some xerophytic adaptations. I would further hypothesize that this flora persisted with extensive modification into our time and is what we today call the flora of the Chaco Dominion. I will call this flora "the Tertiary-Chaco paleo-

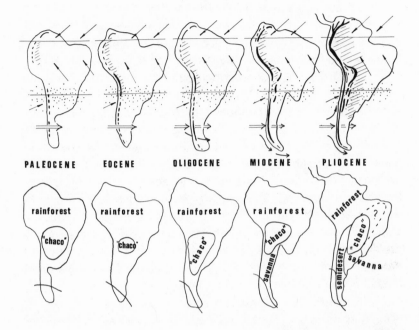

Fig. 2-2. Reconstruction of the outline of South America, mountain chains, and probable vegetation during the Tertiary. Solid line in Patagonia indicates the extent of the Nothofagus forest. Arrows indicate main global wind patterns.

flora" (fig. 2-2). I would like to hypothesize further that *Prosopis*, *Acacia*, and other Mimosoid legumes were elements of that flora as well as taxa in the Anacardiaceae and Zygophyllaceae or their ancestral stock. My justification for this claim is the prominence of these elements in the Chaco Dominion. Another indication of their primitiveness is their present distributional ranges, particularly the fact that many are found in Africa, which was supposedly considerably closer to South America at the beginning of the Tertiary than it is now. Furthermore, fossil remains of *Prosopis*, *Schinopsis*, *Schinus*, *Zygophyllum* (=*Guaiacum* [?], cf. Porter, 1974), and *Aspidosperma* have been reported (Berry, 1930; MacGinitie, 1953, 1969; Kruse, 1954; Axelrod, 1970) from North American floras (Colorado, Utah, and Wyoming) of Eocene or Eocene-Oligocene age (Florissant and Green River). This would indicate a wide distribution for these elements already at the beginning of the Tertiary. Verification of this hypothesis can be obtained from the study of the geology of Maestrichtian and Paleocene deposits of central Argentina and Chile as well as from the study of microfossils (and even megafossils) of this area.

Axelrod (1970) has proposed that some of the genera of the Monte and Chaco that have no close relatives in their respective families, such as *Donatia*, *Puya*, *Grabowskia*, *Monttea*, *Bredemeyera*, *Bulnesia*, and *Zuccagnia*, are modified relicts from the original upland angiosperm stock, which he feels is of pre-Cretaceous origin. The question of a pre-Cretaceous origin for the angiosperms is a speculative point due, to large measure, to the lack of corroborating fossil evidence (Axelrod, 1970). Assuming Axelrod's thesis were correct, not only would the Tertiary-Chaco paleoflora be the ancient nucleus of the subtropical elements adapted to a dry season, but presumably it would also represent the oldest angiosperm stock on the continent. However, some of the genera cited (*Bulnesia*, *Monttea*, *Zuccagnia*) are not primitive but are highly specialized.

The Tertiary in South America is characterized by the gradual lifting of the Andean chain. The process became much accelerated after the Eocene. The Tertiary is also characterized by the gradual cooling and drying of the climate after the Eocene, apparently a world-wide event (Wolfe and Barghoorn, 1960; Axelrod and Bailey, 1969; Wolfe, 1971). As a result, during the Paleocene and Eocene the Tertiary-Chaco paleoflora must have been fairly restricted in its distribution. However, after the Eocene it started to expand and differentiate. During the Eocene the first development of the steppe elements of the pampa region occurred. (The present pampa flora is part of the Chaco Dominion.) Evidence is found in the evolution of mammals adapted to eating grass and living in open habitats (Patterson and Pascual, 1972) and in the first fossil evidence of grasses (Van der Hammen, 1966). Since Tertiary deposits exist in the pampa region that have yielded animal microfossils (Padula, 1972), a study of these cores for plant microfossils may produce uncontroversial direct evidence for the evolution of the pampas. Less certain is the evolution of a dry Chaco or Monte vegetation from the Tertiary-Chaco paleoflora during the latter part of the Tertiary. The present-day distribution of *Larrea*, *Bulnesia*, *Monttea*, *Geoffroea*, *Cercidium*, and other typical dry Chaco or Monte elements makes me think that by the Pliocene a semidesert-type vegetation, not just one adapted to a dry season, was in existence in western middle South America. In effect, all these elements, so characteristic of extensive areas of semidesert in Argentina east of the Andes, are represented by small or, in some cases (*Geoffroea*, *Prosopis*), fairly abundant populations west of the Andes in a region (the Pacific Coastal Desert) dominated today by Andean elements that originated at a later date. Furthermore, the Mediterranean region of Chile is formed in part by Chaco elements, and we know that a Mediterranean climate probably did not evolve until the Pleistocene (Raven, 1971, 1973; Axelrod, 1973). It is, consequently, probable that toward the end of the Tertiary a more xerophytic flora was evolving from the Tertiary-Chaco paleoflora, perhaps as a result of xeric local conditions in the lee of the rising mountains. It should be pointed out, too, that the coastal cordillera in Chile rose first and was bigger in the Tertiary than it is today (Cecioni, 1970).

The Pliocene is characterized by a great increase in orogenic activity that created the Cordillera de los Andes in its present form (Kummel, 1961; Harrington, 1962; Haffer, 1970). In a sea of essentially tropical vegetation an alpine environment was created, ready to be colonized by plants that could withstand not only the cold but also the great daily (in low latitudes) or seasonal climate variation (in high latitudes). The rise of the cordillera also interfered with the free flow of winds and produced changes in local climate, creating the Patagonian and Monte semideserts as we know them today.

Most of the elements that populate the high cordillera were drawn from the Neotropical flora, although some Antarctic elements invaded the open Andean regions (*Lagenophora*) as well as some North American elements, such as *Alnus*, *Sambucus*, *Viburnum*, *Erigeron*, *Aster*, and members of the Ericaceae and Cruciferae (Van der Hammen, 1966). Particularly, elements in the Compositae, Caryophyllaceae, Umbelliferae, Cruciferae, and Gramineae radiated and became dominant in the newly opened habitats. The Puna is an integral part of the Andes and consequently must have become populated by the Andean elements as it became uplifted in the Pliocene, displacing the original tropical Amazonian elements present there in Miocene times (Berry, 1917) which were ill-adapted to the new climatic conditions. A similar but less-evident pattern must have taken place in Patagonia. The tropical flora present in Patagonia in the Paleocene and Eocene and, on mammalian evidence, up to the Miocene (Patterson and Pascual, 1972) had migrated north in response to the cooling climate and had been apparently replaced by an open steppe presumably of Chaco origin. This flora and that of the evolving dry Chaco-Monte vegetation should have been able to adapt to the increasing xerophytic environment of Patagonia in the Pliocene and perhaps did. Only better fossil evidence can tell. But it is the events of the Pleistocene that account for the present dry flora of Patagonia. The same can be said for the Pacific Coastal Desert. The lifting of the cordillera created a disjunct area of Chaco vegetation on the Pacific coast. Before the development of the cold Humboldt Current, presumably a Pleistocene event (in its present condition), the environment must have been more mesic, especially in its southern regions, and the Chaco-Monte flora should have been able to adapt to those conditions. Again it is the Pleistocene events that explain the present flora.

The Pleistocene is marked by a series (up to four) of very drastic fluctuations in temperature as well as some (presumably also up to four) extreme periods of aridity. During the cold periods permanent snow lines dropped, mountain glaciers formed, and the high mountain vegetation expanded. During the last glacial event, the Würm, which seems to have been the most drastic in South America, ice fields extended in Patagonia from the Pacific Coast to some 100 km or more east of the high mountain line, and the Patagonian climate was that of an arctic steppe with extremely cold winters. During those periods any existing subtropical elements disappeared and were replaced by cold-adapted Andean taxa. As conditions gradually improved, phyletic evolution of that cold-adapted flora of Andean origin produced today's Patagonian flora. The same is true in the Pacific Coastal Desert. Only here some elements of the old Chaco flora managed to survive the Pleistocene and account for the range disjunctions of such species as *Larrea divaricata*, *Bougainvillea spinosa*, or *Bulnesia retama*. Some elements of the old Chaco flora of the Pacific coast moved north and are found today in the *algarrobal* formation of northern Peru and Ecuador and in dry inter-Andean valleys of Colombia and Venezuela. Others moved south and are part of the Espinal and Matorral formations of Chile.

The Pleistocene also affected the Monte region. The northern inter-

Andean valleys and *bolsones* became colder, and the mountain slope vegetation was replaced by a vegetation of high-mountain type, so that the Monte vegetation was compressed into a smaller area farther south and east, principally in Mendoza, La Pampa, San Luis, and La Rioja. Cold periods were followed by warmer and wetter periods, characterized by more mesic subtropical elements which expanded their range from the east slope of the Andes to Brazil (Smith, 1962; Simpson Vuilleumier, 1971). After these mesic periods came very extensive dry periods, marked by expansion of the Monte flora and the breakup of the ranges of the subtropical elements, many of which have now disjunct distributions in Brazil and the eastern Andes.

Acknowledgments

This paper is the result of my long-standing interest in the flora and vegetation of the Monte in Argentina and of temperate semidesert and desert regions in general. Too many people to name individually have aided my interest, stimulated my curiosity, and satisfied my knowledge for facts. I would like, however, to acknowledge my particular indebtedness to Professor Angel L. Cabrera at the University of La Plata in Argentina, who first initiated me into floristic studies and who, over the years, has continuously stimulated me through personal conversations and letters, and through his writings. Other people whose help I would like to acknowledge are Drs. Humberto Fabris, Juan Hunziker, Harold Mooney, Jorge Morello, Arturo Ragonese, Beryl Simpson, and Federico Vervoorst. With all of them and many others I have discussed the ideas in this paper, and no doubt these ideas became modified and were changed to the point where it is hard for me to state now exactly what was originally my own. I further would like to thank the Milton Fund of Harvard University, the University of Michigan, and the National Science Foundation, which made possible yearly trips to South America over the last ten years. I particularly would like to acknowledge two NSF grants for studies in the structure of ecosystems that have supported my active research in the Monte and Sonoran desert ecosystem for the last three years. Sergio Archangelsky, Angel L. Cabrera, Philip Cantino, Carlos Menéndez,

Bryan Patterson, Duncan Porter, Beryl Simpson, and Rolla Tryon read the manuscript and made valuable suggestions for which I am grateful.

Summary

The existent evidence regarding the floristic relations of the semidesert regions of South America and how they came to exist is reviewed.

The regions under consideration are the phytogeographical provinces of Patagonia and the Monte in Argentina; the Puna in Argentina, Bolivia, and Peru; the Espinal in Chile; and the Pacific Coastal Desert in Chile and Peru. It is shown that the flora of Patagonia, the Puna, and the Pacific Coastal Desert are basically of Andean affinities, while the flora of the Monte has affinities with the flora of the subtropical Chaco. However, Chaco elements are also found in Chile and Peru. From these considerations and those of a geological and geoclimatological nature, it is postulated that there might have existed an early (Late Cretaceous or early Tertiary) flora adapted to living in more arid —although not desert—environments in and around lat. 30° S.

The present flora of the desert and semidesert temperate regions of South America is largely a reflection of Pleistocene events. The flora of the Andean Dominion that originated the flora that today populates Patagonia and the Pacific Coastal Desert, however, evolved largely in the Pliocene, while the Chaco flora that gave origin to the Monte and Prepuna flora had its beginning probably as far back as the Cretaceous.

References

Ahlfeld, F., and Branisa, L. 1960. *Geología de Bolivia*. La Paz: Instituto Boliviano de Petroleo.
Auer, V. 1958. The Pleistocene of Fuego Patagonia. II. The history of the flora and vegetation. *Suomal. Tiedeakat. Toim. Ser. A 3.* 50:1–239.

————. 1960. The Quaternary history of Fuego-Patagonia. *Proc. R. Soc. Ser. B.* 152:507–516.

Axelrod, D. I. 1967. Drought, diastrophism and quantum evolution. *Evolution* 21:201–209.

————. 1970. Mesozoic paleogeography and early angiosperm history. *Bot. Rev.* 36:277–319.

————. 1973. History of the Mediterranean ecosystem in California. In *The convergence in structure of ecosystems in Mediterranean climates*, ed. H. Mooney and F. di Castri, pp. 225–284. Berlin: Springer.

Axelrod, D. I., and Bailey, H. P. 1969. Paleotemperature analysis of Tertiary floras. *Paleogeography, Paleoclimatol. Paleoecol.* 6:163–195.

Barbour, M. G., and Díaz, D. V. 1972. *Larrea* plant communities on bajada and moisture gradients in the United States and Argentina. *U.S./Intern. biol. Progr.: Origin and Structure of Ecosystems Tech. Rep.* 72–6:1–27.

Bartlett, A. S., and Barghoorn, E. S. 1973. Phytogeographic history of the Isthmus of Panama during the past 12,000 years. In *Vegetation and vegetational history of northern Latin America*, ed. A. Graham, pp. 203–300. Amsterdam: Elsevier.

Berry, E. W. 1917. Fossil plants from Bolivia and their bearing upon the age of the uplift of the eastern Andes. *Proc. U.S. natn. Mus.* 54:103–164.

————. 1930. Revision of the lower Eocene Wilcox flora of the southeastern United States. *Prof. Pap. U.S. geol. Surv.* 156:1–196.

Böcher, T.; Hjerting, J. P.; and Rahn, K. 1963. Botanical studies in the Atuel Valley area, Mendoza Province, Argentina. *Dansk bot. Ark.* 22:7–115.

Burkart, A. 1940. Materiales para una monografía del género *Prosopis*. *Darwiniana* 4:57–128.

————. 1952. *Las leguminosas argentinas silvestres y cultivadas*. Buenos Aires: Acme Agency.

Cabrera, A. L. 1947. La Estepa Patagónica. In *Geografía de la República Argentina*, ed. GAEA, 8:249–273. Buenos Aires: GAEA.

————. 1953. Esquema fitogeográfico de la República Argentina. *Revta Mus. La Plata (nueva Serie), Bot.* 8:87–168.

————. 1958. La vegetación de la Puna Argentina. *Revta Invest. agríc., B. Aires* 11:317–412.

————. 1971. Fitogeografía de la República Argentina. *Boln Soc. argent. Bot.* 14:1–42.

Caldenius, C. C. 1932. Las glaciaciones cuaternarias en la Patagonia y Tierra del Fuego. *Geogr. Annlr* 14:1–164.

Castellanos, A. 1956. Caracteres del pleistoceno en la Argentina. *Proc.IV Conf. int. Ass. quatern. Res.* 2:942–948.

Cecioni, G. 1970. *Esquema de paleogeografía chilena*. Santiago: Editorial Universitaria.

Charlesworth, J. K. 1957. *The Quaternary era.* 2 vols. London: Arnold.

Czajka, W. 1966. Tehuelche pebbles and extra-Andean glaciation in east Patagonia. *Quaternaria* 8:245–252.

Czajka, W., and Vervoorst, F. 1956. Die naturräumliche Gliederung Nordwest-Argentiniens. *Petermanns geogr. Mitt.* 100:89–102, 196–208.

Darlington, P. J. 1957. *Zoogeography: The geographical distribution of animals*. New York: Wiley.

————. 1965. *Biogeography of the southern end of the world*. Cambridge, Mass.: Harvard Univ. Press.

Dietz, R. S., and Holden, J. C. 1970. Reconstruction of Pangea: Breakup and dispersion of continents, Permian to present. *J. geophys. Res.* 75:4939–4956.

Dimitri, M. J. 1972. Consideraciones sobre la determinación de la superficie y los limites naturales de la región andino-patagónica. In *La región de los Bosques Andino-Patagonicos*, ed. M. J. Dimitri, 10:59–80. Buenos Aires: Col. Cient. del INTA.

Emiliani, C. 1966. Isotopic paleotemperatures. *Science, N.Y.* 154: 021.

Ferreyra, R. 1960. Algunos aspectos fitogeográficos del Perú. *Publnes Inst. Geogr. Univ. San Marcos (Lima)* 1(3):41–88.

Feruglio, E. 1949. *Descripción geológica de la Patagonia*. 2 vols. Buenos Aires: Dir. Gen. de Y.P. F.

Fiebrig, C. 1933. Ensayo fitogeográfico sobre el Chaco Boreal. *Revta Jard. bot. Mus. Hist. nat. Parag.* 3:1–87.

Flint, R. F., and Fidalgo, F. 1968. Glacial geology of the east flank of the Argentine Andes between latitude 39°-10′ S and latitude 41°-20′ S. *Bull. geol. Soc. Am.* 75:335–352.

Flohn, H. 1969. *Climate and weather*. New York: McGraw-Hill Book Co.

Frenguelli, J. 1957. El hielo austral extraandino. In *Geografía de la República Argentina*, ed. GAEA, 2:168–196. Buenos Aires: GAEA.

Frenzel, B. 1968. The Pleistocene vegetation of northern Eurasia. *Science, N.Y.* 161:637.

Good, R. 1953. *The geography of the flowering plants*. London: Longmans, Green & Co.

Goodspeed, T. 1945. The vegetation and plant resources of Chile. In *Plants and plant science in Latin America*, ed. F. Verdoorn, pp. 147–149. Waltham, Mass.: Chronica Botanica.

Gordillo, C. E., and Lencinas, A. N. 1972. Sierras pampeanas de Córdoba y San Luis. In *Geología regional Argentina*, ed. A. F. Leanza, pp. 1–39. Córdoba: Acad. Nac. de Ciencias.

Groeber, P. 1936. Oscilaciones del clima en la Argentina desde el Plioceno. *Revta Cent. Estud. Doct. Cienc. nat., B. Aires* 1(2):71–84.

Haffer, J. 1969. Speciation in Amazonian forest birds. *Science, N.Y.* 165:131–137.

————. 1970. Geologic-climatic history and zoogeographic significance of the Uraba region in northwestern Colombia. *Caldasia* 10: 603–636.

Harrington, H. J. 1962. Paleogeographic development of South America. *Bull. Am. Ass. Petrol. Geol.* 46:1773–1814.

Haumann, L. 1947. Provincia del Monte. In *Geografía de la República Argentina*, ed. GAEA, 8:208–248. Buenos Aires: GAEA.

Hecht, M. K. 1963. A reevaluation of the early history of the frogs. Pt. II. *Syst. Zool.* 12:20–35.

Hoffstetter, R. 1972. Relationships, origins and history of the Ceboid monkeys and caviomorph rodents: A modern reinterpretation. *Evol. Biol.* 6:323–347.

Hünicken, M. 1966. Flora terciaria de los estratos del río Turbio, Santa Cruz. *Revta Fac. Cienc. exact. fís. nat. Univ. Córdoba, Ser. Cienc. nat.* 27:139–227.

Hunziker, J. H.; Palacios, R. A.; de Valesi, A. G.; and Poggio, L. 1973. Species disjunctions in *Larrea*: Evidence from morphology, cytogenetics, phenolic compounds, and seed albumins. *Ann. Mo. bot. Gdn* 59:224–233.

Johnston, I. 1924. Taxonomic records concerning American spermatophytes. 1. Parkinsonia and Cercidium. *Contr. Gray Herb. Harv.* 70:61–68.

———. 1929. Papers on the flora of northern Chile. *Contr. Gray Herb. Harv.* 85:1–171.

Kruse, H. O. 1954. Some Eocene dicotyledoneous woods from Eden Valley, Wyoming. *Ohio J. Sci.* 54:243–267.

Kummel, B. 1961. *History of the earth*. San Francisco: W. H. Freeman & Co.

Liddle, R. A. 1946. *The geology of Venezuela and Trinidad.* 2d ed. Ithaca: Pal. Res. Inst.

Lorentz, P. 1876. Cuadro de la vegetación de la República Argentina. In *La República Argentina*, ed. R. Napp, pp. 77–136. Buenos Aires: Currier de la Plata.

MacGinitie, H. D. 1953. Fossil plants of the Florissant beds, Colorado. *Publs Carnegie Instn* 599:1–180.

———. 1969. The Eocene Green River flora of northwestern Colorado and northeastern Utah. *Univ. Calif. Publs geol. Sci.* 83:1–140.

Mayr, E. 1963. *Animal species and evolution*. Cambridge, Mass: Harvard Univ. Press.

Menéndez, C. A. 1969. Die fossilen floren Südamerikas. In *Biogeography and ecology in South America*, ed. E. J. Fittkau, J. Illies, H. Klinge, G. H. Schwabe, and H. Sioli, 2:519–561. The Hague: Dr. W. Junk.

———. 1972. Paleofloras de la Patagonia. In *La región de los Bosques Andino-Patagonicos*, ed. M. J. Dimitri, 10:129–184. Col. Cient. Buenos Aires: del INTA.

Mooney, H. and Dunn, E. L. 1970. Convergent evolution of Mediterranean climate evergreen sclerophyll shrubs. *Evolution* 24:292–303.

Morello, J. 1958. La provincia fitogeográfica del Monte. *Op. lilloana* 2:1–155.

Padula, E. L. 1972. Subsuelo de la mesopotamia y regiones adyacentes. In *Geología regional Argentina*, ed. A. F. Leanza, pp. 213–236. Córdoba: Acad. Nac. de Ciencias.

Pantin, C. F. A. 1960. A discussion on the biology of the southern cold temperate zone. *Proc. R. Soc. Ser. B.* 152:431–682.

Patterson, B., and Pascual, R. 1972. The fossil mammal fauna of South America. In *Evolution, mammals, and southern continents*, ed. A. Keast, F. C. Erk, and B. Glass, pp. 247–309. Albany: State Univ. of N.Y.

Petriella, B. 1972. Estudio de maderas petrificadas del Terciario inferior del área de Chubut Central. *Revta Mus. La Plata (Nueva Serie), Pal.* 6:159–254.

Porter, D. M. 1974. Disjunct distributions in the New World Zygophyllaceae. *Taxon* 23:339–346.

Raven, P. H. 1971. The relationships between "Mediterranean" floras. In *Plant life of South-West Asia*, ed. P. H. Davis, P. C. Harper, and I. C. Hedge, pp. 119–134. Edinburgh: Bot. Soc.

———. 1973. The evolution of Mediterranean floras. In *The convergence in structure of ecosystems in Mediterranean climates*, ed. H. Mooney and F. di Castri, pp. 213–224. Berlin: Springer.

Reiche, K. 1934. *Geografía botánica de Chile*. 2 vols. Santiago: Imprenta Universitaria.

Romero, E. 1973. Ph.D. dissertation, Museo La Plata Argentina.

Sarmiento, G. 1972. Ecological and floristic convergences between seasonal plant formations of tropical and subtropical South America. *J. Ecol.* 60:367–410.

Schuchert, C. 1935. *Historical geology of the Antillean-Caribbean region*. New York: Wiley.

Schwarzenbach, M. 1968. Das Klima des rheinischen Tertiärs. *Z. dt. geol. Ges.* 118:33–68.

Simpson, G. G. 1950. History of the fauna of Latin America. *Am. Scient.* 1950:361–389.

Simpson Vuilleumier, B. 1967. The systematics of Perezia, section Perezia (Compositae). Ph.D. thesis, Harvard University.

———. 1971. Pleistocene changes in the fauna and flora of South America. *Science, N.Y.* 173:771–780.

Smith, L. B. 1962. Origins of the flora of southern Brazil. *Contr. U.S. natn. Herb.* 35:215–250.

Solbrig, O. T. 1972. New approaches to the study of disjunctions with special emphasis on the American amphitropical desert disjunctions. In *Taxonomy, phytogeography and evolution*, ed. D. D. Valentine, pp. 85–100. London and New York: Academic Press.

———. 1973. The floristic disjunctions between the "Monte" in Argentina and the "Sonoran Desert" in Mexico and the United States. *Ann. Mo. bot. Gdn* 59:218–223.

Soriano, A. 1949. El limite entre las provincias botánicas Patagónica y Central en el territorio del Chubut. *Revta argent. Agron.* 17:30–66.

———. 1950. La vegetación del Chubut. *Revta argent. Agron.* 17:30–66.

———. 1956. Los distritos floristicos de la Provincia Patagónica. *Revta Invest. agríc., B. Aires* 10:323–347.

Stebbins, G. L. 1950. *Variation and evolution in plants.* New York: Columbia Univ. Press.

———. 1952. Aridity as a stimulus to evolution. *Am. Nat.* 86:33–44.

Steinmann, G. 1930. *Geología del Perú.* Heidelberg: Winters.

Thorne, R. F. 1973. Floristic relationships between tropical Africa and tropical America. In *Tropical forest ecosystems in Africa and South America: A comparative review*, ed. B. J. Meggers, E. S. Ayensu, and D. Duckworth, pp. 27–40. Washington, D.C.: Smithsonian Instn. Press.

Van der Hammen, T. 1961. The Quaternary climatic changes of northern South America. *Ann. N.Y. Acad. Sci.* 95:676–683.

———. 1966. Historia de la vegetación y el medio ambiente del norte sudamericano. In *1° Congr. Sud. de Botánica, Memorias de Symposio*, pp. 119–134. Mexico City: Sociedad Botánica de Mexico.

Van der Hammen, T., and González, E. 1960. Upper Pleistocene and Holocene climate and vegetation of the "Sabana de Bogotá." *Leid. geol. Meded.* 25:262–315.

Vanzolini, P. E., and Williams, E. E. 1970. South American anoles: The geographic differentiation and evolution of the *Anolis chrysolepis* species group (Sauria, Iguanidae). *Archos Zool. Est. S Paulo* 19:1–298.

Vervoorst, F. 1945. *El Bosque de algarrobos de Pipanaco (Catamarca).* Ph.D. dissertation, Universidad de Buenos Aires

———. 1973. Plant communities in the Indian de Pipanaco. U.U./ Intern. biol. Progr.: Origin and Structure of Ecosystems Prog. Rep. 73-3:3–17.

Volkheimer, W. 1971. Aspectos paleoclimatológicos del Terciario Argentina. *Revta Mus. Cienc. nat. B. Rivadavia Paleontol.* 1:243–262.

Vuilleumier, F. 1967. Phyletic evolution in modern birds of the Patagonian forests. *Nature, Lond.* 215:247–248.

Weberbauer, A. 1945. *El mundo vegetal de los Andes Peruanos*. Lima: Est. Exp. La Molina.

Wijmstra, T. A., and Van der Hammen, T. 1966. Palynological data on the history of tropical savannas in northern South America. *Leid. geol. Meded.* 38:71–90.

Wolfe, J. A. 1971. Tertiary climatic fluctuations and methods of analysis of Tertiary floras. *Paleogeography, Paleoclimatol. Paleoecol.* 9:27–57.

Wolfe, J. A., and Barghoorn, E. S. 1960. Generic change in Tertiary floras in relation to age. *Am. J. Sci.* 258A:388–399.

Wright, H. E., and Frey, D. G. 1965. *The Quaternary of the United States*. Princeton: Princeton Univ. Press.

Yang, T. W. 1970. Major chromosome races of *Larrea divaricata* in North America. *J. Ariz. Acad. Sci.* 6:41–45.

3. The Evolution of Australian Desert Plants   John S. Beard

Introduction

As an opening to this subject it may be well to outline briefly the where-abouts of the Australian desert, its climate and vegetation. The desert consists, of course, of the famous "dead heart" of Australia, covering the interior of the continent; and it has been defined on a map together with its component natural regions by Pianka (1969a). An important characteristic of this area is that, while certainly arid and classifiable as desert by most, if not all, of the better-known bioclimatic classifi-cations and indices, it is not as rainless as some of the world's deserts and is correspondingly better vegetated. The most arid portion of the Australian interior, the Simpson Desert, receives an average rainfall of 100 mm, while most of the rest of the desert receives around 200 mm. The desert is usually taken to begin, in the south, at the 10-inch, or 250-mm, isohyet. In the north, in the tropics under higher temper-atures, desert vegetation reaches the 20-inch, or 500-mm, isohyet.

Plant Formations in Australian Deserts

As a result of the rainfall in the Australian desert, it always possesses a plant cover of some kind, and we have no bare and mobile sand dunes and few sheets of barren rock. There are two principal plant formations: a low woodland of *Acacia* trees colloquially known as mul-ga, which covers roughly the southern half of the desert south of the tropic, and the "hummock grassland" (Beadle and Costin, 1952) col-loquially known as spinifex, which covers the northern half within the tropics. Broadly the two formations are climatically separated, although the preference of each of them for certain soils tends to obscure this relationship; thus, the hummock grassland appears on sand even in the southern half. The *Acacia* woodland is to be compared with those of other continents, but few Australian species of *Acacia* have thorns and few have bipinnate leaves. The hummock grassland, on the other hand, is, I think, a unique product of evolution in Australia. It is comparable with the grass steppe vegetation of other continents, but the life form of the grasses is different. Two genera are represented, *Triodia* and *Plectrachne*. Each plant branches repeatedly into a great number of culms which intertwine to form a hummock and bear rigid, terete, pungent leaves presenting a serried phalanx to the exterior. When flowering takes place in the second half of summer, given adequate rains, upright rigid inflorescences are produced above the crown of the hummock, rising 0.5 to 1 m above it. The flowers quickly set seed, which is shed within two months, although this is then the beginning of the dry season. The size of the hummock varies considerably according to the site from 30 cm in height and diameter on the poorest, stoniest sites up to about 1 m in height and 2 m in diameter on some deep sands. Old hummocks, if unburnt, tend to die out in the center or on one side, leading to ring or crescentic growth. At this stage the original root has died and the outer culms have rooted themselves adventitiously in the soil. Individual hummocks do not touch, and there is much bare ground between them.

The hummock grassland normally contains a number of scattered shrubs or scattered trees in less-arid areas where ground water is available. All of these must be resistant to fire, by which the grassland is regularly swept. After burning, the grasses regenerate from the root or from seed.

The *Acacia* woodlands, in which *A. aneura* is frequently the sole species in the upper stratum, contain a sparse lower layer of shrubs most frequently of the genera *Eremophila* and *Cassia*, 1–2 m tall, and an even sparser ground layer mainly of ephemerals and only locally of grasses.

These Australian desert formations are given distinctive character by the physiognomy of their commonest plants, that is:

Trees. Evergreen, sclerophyll. Leaves pendent in *Eucalyptus*; linear, erect, and glaucescent in *Acacia aneura*; vestigial in *Casuarina decaisneana*. Bark white in most species of *Eucalyptus*.

Shrubs. The larger shrubs are sclerophyll, typically phyllodal species of *Acacia*; the smaller shrubs, ericoid (*Thryptomene*).

Subshrubs. Many soft perennial subshrubs typically with densely pubescent or silver-tomentose stems and leaves, e.g., *Crotalaria cunninghamii*, and numerous Verbenaceae (*Dicrastyles, Newcastelia, Pityrodia* spp.). Also, suffrutices with underground rootstocks and ephemeral or more or less perennial shoots, often also densely pubescent or silver-tomentose, e.g., *Brachysema chambersii*, many *Ptilotus* spp., *Leschenaultia helmsii*, and *L. striata*. Some are viscid—*Goodenia azurea* and *G. stapfiana*.

Ephemerals. Many species of Compositae, *Ptilotus*, and *Goodenia* appear as brilliant-flowering annuals in season. Colors are predominantly yellow and mauve, with some white and pink. Red is absent.

Grasses. Grasses of the "short bunch-grass" type in the sense of Bews (1929) occur only on alluvial flats close to creeks or on plains of limited extent developed on or close to basic rocks. In these cases there is a fine soil with a relatively high water-holding capacity and probably also high-nutrient status. On sand, laterite, and rock in the desert, grasses belong almost entirely to the genera *Triodia* and *Plectrachne*, which adopt the hummock-grass form as previously described. This growth form appears to be peculiar to Australia and to be the only unique form evolved in the Australian desert.

It will therefore be seen that the Australian desert possesses special vegetative characters of its own which can be supposed to be of some adaptive significance, particularly *glaucescence* of bark and leaves, *pubescence* frequently in association with glaucescence, *suffrutescence*, the presence of vernicose and viscid leaf surfaces, and the *spinifex* habit in grasses. Other characters, such as tree and shrub growth forms and sclerophylly, are not peculiar to the desert Eremaea but are shared with other Australian vegetation.

Growth Forms

In most of the world's deserts special and peculiar growth forms have evolved which confer advantage in the arid environment. In North and

Central America the family Cactaceae has produced the well-known range of forms based on stem succulence, closely replicated by the Euphorbiaceae in Africa. In southern African deserts leaf succulence is a dominant feature that has been developed in many families, notably the Aizoaceae and Liliaceae. Leaf-succulent rosette plants in the Bromeliaceae are a feature of both arid northwest Brazil and the cold Andean Puna. In all cases we are accustomed to look also for deciduous, thorny trees and plants with underground perennating organs, especially bulbs and corms. In Australia there is an extraordinary lack of all these forms; where some of them exist they are confined to certain areas.

Leaf- and stem-succulent plants belonging to the family Chenopodiaceae in fact characterize two other important plant formations, less widespread than the principal formations described above and confined to certain soils. These I have named "succulent steppe" (Beard, 1969) following the usage of African ecologists; they comprise, first, saltbush and bluebush steppe dominated by species of *Atriplex* and *Kochia* respectively, and, second, samphire communities with *Arthrocnemum*, *Tecticornia*, and related genera. The former are small soft shrubs whose leaves are fleshy or semisucculent, associated with annual grasses and herbs, and sometimes with a sclerophyll tree layer of *Acacia* or *Eucalyptus*. The formation is confined to the southern half of the desert region and occupies alkaline soils, most commonly on limestone or calcareous clays. In the northern half such soils normally carry hummock grassland on limestone and bunch grassland on clays. The samphire communities, however, range throughout the region on very saline soils in depressions, usually in the beds of playa lakes or peripheral to them. The samphires are subshrubs with succulent-jointed stems. These formations are the only ones with a definitely succulent character and are essentially halophytes.

On the siliceous soils of the desert, sclerophylly is the dominant characteristic, and stem succulence is represented in only a handful of species of no prominence, such as *Sarcostemma australe* (Asclepiadaceae), a divaricate, leafless plant found occasionally in rocky places. Others are *Spartothamnella teucriiflora* (Verbenaceae) and *Calycopeplus helmsii* (Euphorbiaceae). Likewise, leaf succulence is found in a variety of groups but is often weakly developed and never a conspicuous feature. *Gyrostemon ramulosus* (Phytolaccaceae) has somewhat fleshy foliage, which the explorers noted as a favorite feed of camels. The Aizoaceae in Australia are mostly tropical herbs, and the most genuinely succulent member, *Carpobrotus*, is not Eremaean. The Portulacaceae are a substantial group with twenty-seven species in *Calandrinia*, of which about twelve are Eremaean, and eight in *Portulaca*, which belong to the Northern Province. *Calandrinia* is herbaceous and leaf succulent, and several species are not uncommon, but it will be noted that they are not essentially desert plants. A weak leaf succulence can be seen in *Kallstroemia, Tribulus*, and *Zygophyllum* of the Zygophyllaceae and in *Euphorbia* and *Phyllanthus* of the Euphorbiaceae. Few of these are plants of any ecological importance.

Evolutionary History

The evolutionary significance of these different growth forms must now be discussed. Our view of the past history of biota has been transformed by the development of the theory of plate tectonics in quite recent years, with sanction given to the previously heretical ideas of continental drift. As long ago as 1856, in his famous preface to the *Flora Tasmaniae*, J. D. Hooker suggested that the modern Australian flora was compounded of three elements—an Indo-Malaysian element derived from southeast Asia, an autochthonous element evolved within Australia itself, and an Antarctic element comprising forms common to the southern continents which in some way should be presumed to have been transmitted via Antarctica. The trouble was that, while the reality of this Antarctic element could not be doubted, no means or mechanism save that of long-range dispersal could be used to account for it—unless one were very daring and, after Wegener and du Toit, were prepared to invoke continental drift. The thinking of those years of fixed-positional geology is typified by Darlington's book *Biogeography of the Southern End of the World* (1965), in which the southern continents are seen as refuges where throughout time odd forms from the Northern Hemisphere have established themselves and survived. Our Antarctic element would then become only a random selection of forms long extinct in the other hemisphere. This view is now discredited.

Although the breakup of Gondwanaland is dated rather earlier than the origin of the angiosperms, many of the continents do seem to have remained sufficiently close or, in some cases, in actual contact in such a way that explanations of the distribution of plant forms are materially assisted. Where Australia is concerned in this discussion of desert biota we need only go back to Eocene times, some 40 to 60 million years ago, when our continent was joined to Antarctica along the southern edge of its continental shelf and lay some 15° of latitude farther south than now (Griffiths, 1971). In middle Eocene times a rift occurred in the position of the present mid-oceanic ridge separating Australia and Antarctica; the two continents broke apart and drifted in opposite directions: Antarctica to have its biota largely extinguished by a polar icecap, Australia to move toward and into the tropics, passing in the process through an arid zone in which much of it still lies. The evolution of the desert flora of Australia has therefore occurred since the Eocene *pari passu* with this movement.

In discussions of Tertiary paleoclimates it is commonly assumed that the circulation of the atmosphere has always been much the same as it is today, so that the positions of major latitudinal climatic belts have also been fairly constant, even though there may have been cyclic variations in temperature and in quantity of rainfall. At the time, therefore, when Australia was situated 15° farther south, it would have lain squarely in the roaring forties; and it seems likely that a copious and well-distributed rainfall would have been received more or less throughout the continent. This is borne out by the fossil record which predominantly suggests a cover of rain forest of a character and composition similar to that found today in the North Island of New Zealand (Raven, 1972).

Paleontological evidence suggests rather warmer temperatures prevailing at that time and in those latitudes than exist there today. When the break from Antarctica took place, the southern coastline of Australia slumped and thin deposits of Eocene and Miocene sediments were laid down upon the continental margin. Fossils indicate deposition in seas of tropical temperature, continuing as late as Miocene times (Dorman, 1966; Cockbain, 1967; Lowry, 1970). This is consistent with the evidence of tropical flora extending to lat. 50° N in North America in the Eocene (Chaney, 1947) and to Chile and Patagonia (Skottsberg, 1956).

Evidence from the soil supports the concept of both high temperature and high rainfall. In the Canning and Officer sedimentary basins in Western Australia, the parts of the country now occupied by the Great Sandy and Gibson deserts, an outcrop of rocks of Cretaceous age has been very deeply weathered and thickly encrusted with laterite. Farther south than this an outcrop of Miocene limestone in the Eucla basin exhibits relatively little weathering or development of typical karst features and is considered to have been exposed to a climate not substantially wetter than the present since its uplift from the sea at the end of Miocene times (Jennings, 1967).

The laterization would indicate subjection for a long period to a warm, wet climate, which must therefore be early Tertiary in date. The present surface features of all of these sedimentary basins are in accord with presumed climatic history based on known latitudinal movement of the continent.

From Eocene times, therefore, Australian flora had to adapt itself to progressive desiccation. It is frequently assumed that it also had to adapt to warmer temperatures in moving northward, but I believe that this is a mistake. We have fossil evidence for warmer temperatures already in the Eocene, followed by a progressive cooling of the earth through the later Tertiary; and the northward movement of Australasia largely provided, I think, a compensation for the latter process. I do not concur with Raven and Axelrod (1972), for example, that we have to assume a developed adaptation to tropical conditions in those elements in the flora of New Caledonia which are of southern origin. Australasian flora, however, had to adapt to the greater extremes of temperature which accompany aridity, even though mean temperatures may not have greatly altered.

From my own consideration of the paleolatitudes and an attempt to map the probable paleoclimates (which I cannot now go into in detail), I believe that the first appearance of aridity may have been in the northwest in the Kimberley district of Western Australia in later Eocene times, expanding steadily to the southeast. The first Mediterranean climate with its winter-wet, summer-dry regime seems likely to have become established in the Pilbara district of Western Australia in the Oligocene and to have been progressively displaced southward.

The Roles of Fires and Soil

In addition to the climatic adaptations required, Australian flora also had to adapt itself to changes in soil which have accompanied the desiccation and to withstand fire. In the early Tertiary rain forests fire was probably unknown or a rarity. Such forests are able to grow even on a highly leached and impoverished substratum in the absence of fire, as a cycle of accumulation and decomposition of organic matter is built up and the forest is living on the products of its own decay. It has been shown that intense weathering and laterization occurred in the early Tertiary in some areas of Western Australia, and this may be observed elsewhere in the continent.

This process would have occurred initially under the forest without provoking significant changes, but with desiccation two things happen: fire ruptures the nutrient cycle leading to a collapse of the ecosystem, and the laterites are indurated to duricrust. After burning and rapid removal of mineralized nutrients by the wind and the rain, a depauperate scrub community with a low-nutrient demand replaces the rain forest. In the absence of fire a slow succession back to the rain forest will ensue, but further fires stabilize the disclimax. This process may be seen in operation today in western Tasmania. It is intensified where laterite is present since induration of laterite by desiccation is irreversible and produces an inhospitable hardpan in the soil, usually followed by denudation at the least hint early topsoil to leave a surface duricrust which is even more inhospitable.

Arid Australia is situated in those central and western parts of the continent where there has been little or no tectonic movement during the Tertiary to regenerate systems of erosion, so that after desiccation set in there was mostly no widespread removal of ancient weathered soil material or the rejuvenation of the soils. Great expanses of inert sand or surface laterite clothe the higher ground and offer an inhospitable substratum to plants, poor in nutrients and in water-holding capacity. Leaching has continued, and its products have been deposited in the lower ground by evaporation where soils have been zonally accumulating calcium carbonate, gypsum, and chlorides.

Biogeographical Elements

Evolutionary adaptation to these changed conditions during the later Tertiary produced the autochthonous element in the Australian flora mostly by adaptation of forms present in the previously dominant Antarctic element. The Indo-Malaysian element is a relatively recent arrival and, as may be expected, has colonized mainly the moister tropical habitats. It has not contributed very significantly to the desert biota, but there are a few species whose very names betray their origin in that direction: *Trichodesma zeylanicum*, an annual herb in the Boraginaceae; *Crinum asiaticum*, a bulbous Amaryllid, bringing a life form (the perennating bulb) which is almost unknown in Australian desert biota in spite of its apparent evolutionary advantages.

Herbert (1950) pointed out that the autochthonous element is essentially one adapted to subhumid, semiarid, and desert conditions which has been evolved within Australia from forms whose relatives are of world-wide distribution. Evolution, said Herbert, took place in three ways: from ancestors already adapted to these drier climates, by survival of hardier types when increasing aridity drove back the more mesic vegetation, and by recolonization of drier areas by the more xerophytic members of mesic communities.

Burbidge (1960) examined the question more closely and acknowledged a suggestion made to her by Professor Smith-White of the University of Sydney that many of the elements in the desert flora may have developed from species associated with coastal habitats. Burbidge considered that such an opinion was supported by the number of genera in the desert flora of Australia which elsewhere in the world are associated with coastal areas, sand dunes, and habitats of saline type. It is certainly a very reasonable assumption that, in a well-watered early Tertiary continent, source material for future desert plants should lie in the flora of the littoral already adapted to drying winds, sand or rock as a substratum, or salt-marsh conditions. Burbidge went on to say that it is not until the late Pleistocene or early

Recent that there is any real evidence in the fossil record for the existence of a desert flora. However, this does not prove it was not there, and the evidence for the northward movement of Australia into the arid zone suggests strongly that it must have begun its evolution at least as early as the Miocene. Pianka (1969b) in discussing Australian desert lizards found that the species density was too great to permit evolution proceeding only from the sub-Recent. An identical argument is bound to apply to flora also. Speciation is too great and too diversified to have originated so recently.

Morphological Evolution

In addition to the systematic evolution of the desert flora, we may usefully discuss also its morphological evolution. It has been shown that some of the life forms considered most typical of desert biota in other continents are inconspicuous or lacking in Australia, for example, deciduousness, spinescence, and underground perennating organs. Other life forms, especially succulence, are limited to particular areas. Morphologically, there is a dualism in Australian desert flora. The typical plant forms of poor, leached siliceous soils are radically different from those of the base-rich alkaline and saline soils. The former are essentially sclerophyllous in the particular manner of so many Australian plant forms from all over the continent which are not confined to the desert. There has even been the evolution of a unique form of sclerophyll grass, the spinifex or hummock-grass form. On the other hand, succulent and semisucculent leaves replace the sclerophyll on base-rich soils. It is evident that aridity alone is not responsible for sclerophylly in Australian plants as has so often been thought. This evidence seems strongly to support the views of Professor N. C. W. Beadle, expressed in numerous papers (e.g., Beadle, 1954, 1966). Beadle has argued for a relationship between sclerophylly and nutrient deficiency, especially lack of soil phosphate. It certainly seems true to say that the plant forms of nutrient-deficient soils in the Australian desert have had the directions of their evolution dictated not only by aridity but by soil conditions as well, soil conditions largely peculiar to Australia as a continent so that this section of the Australian desert flora has acquired a unique character. It has evolved, we may say, within a straitjacket of sclerophylly. This limitation, however, has not been imposed on the ion-accumulating bottom-land soils where plant forms more similar to those of deserts in other continents have evolved.

To look back to what has been said about the taxonomic evolution of the desert flora, limitations are also imposed by the nature of the genetic source material. A subtropical and warm temperate rain forest is not a very promising source area for forms which will have the necessary genetic plasticity for adaptation to great extremes of temperature and aridity, as well as to extremes of soil deficiency. Certain Australian plant families have possessed this faculty, especially the Proteaceae, and this has resulted in a proliferation of highly specialized and adapted species in a relatively limited number of genera. This phenomenon is remarked especially on the soils which have the most extreme nutrient deficiencies or imbalances under widely differing climatic conditions, notably on the Western Australian sand plains, the Hawkesbury sandstone of New South Wales, and the serpentine outcrops in the mountains of New Caledonia, in all of which different species belonging to the same or related Australian genera can be seen forming a similar maquis or sclerophyll scrub. The sclerophyll desert flora has drawn heavily upon this source material, while the nonsclerophyll flora has been influenced particularly by the ability of the family Chenopodiaceae to produce forms adaptable to the particular conditions.

Summary

The Australian desert, covering the interior of the continent, receives an average rainfall of 100 to 250 mm annually and is well vegetated. There are two principal plant formations, *Acacia* low woodland and *Triodia-Plectrachne* hummock grassland, characteristic broadly of the sectors south and north of the Tropic of Capricorn. Component species are typically sclerophyll in form, even the grasses. Non-sclerophyll vegetation of succulent and semisucculent subshrubs locally occupies alkaline soils, in depressions or on limestone and calcareous clays. There is otherwise a notable absence of such xerophytic life forms as stem and leaf succulents, rosette plants, deciduous thorny trees, and plants with bulbs and corms.

Australian desert flora evolved gradually from the end of Eocene times as the continent moved northward into arid latitudes. As the previous vegetation was mainly a subtropical rain forest, it has been suggested that the source material for this evolution came largely from the littoral and seashore. Species had to adapt not only to aridity but also to soils deeply impoverished by weathering under previous humid conditions and not rejuvenated. It is believed that the siliceous, nutrient-deficient soils have been responsible for the predominantly sclerophyllous pattern of evolution; succulence has only developed on base-rich soils.

References

Beadle, N. C. W. 1954. Soil phosphate and the delimitation of plant communities in eastern Australia. *Ecology* 25:370–374.

———. 1966. Soil phosphate and its role in moulding segments of the Australian flora and vegetation with special reference to xeromorphy and sclerophylly. *Ecology* 47:991–1007.

Beadle, N. C. W., and Costin, A. B. 1952. Ecological classification and nomenclature. *Proc. Linn. Soc. N.S.W.* 77:61–82.

Beard, J. S. 1969. The natural regions of the deserts of Western Australia. *J. Ecol.* 57:677–711.

Bews, J. W. 1929. *The world's grasses*. London: Longmans, Green & Co.

Burbidge, N. T. 1960. The phytogeography of the Australian region. *Aust. J. Bot.* 8:75–211.

Chaney, R. W. 1947. Tertiary centres and migration routes. *Ecol. Monogr.* 17:141–148.

Cockbain, A. E. 1967. Asterocyclina from the Plantagenet beds near Esperance, W.A. *Aust. J. Sci.* 30:68.

Darlington, P. J. 1965. *Biogeography of the southern end of the world*. Cambridge, Mass.: Harvard Univ. Press.

Dorman, F. H. 1966. Australian Tertiary paleotemperatures. *J. Geol.* 74:49–61.

Griffiths, J. R. 1971. Reconstruction of the south-west Pacific margin of Gondwanaland. *Nature, Lond.* 234:203–207.

Herbert, D. A. 1950. Present day distribution and the geological past. *Victorian Nat.* 66:227–232.

Hooker, J. D. 1856. Introductory Essay. In *Botany of the Antarctic Expedition, vol. III flora Tasmaniae*, pp. xxvii–cxii.

Jennings, J. N. 1967. Some karst areas of Australia. In *Land form studies from Australia and New Guinea*, ed. J. N. Jennings and J. A. Mabbutt. Canberra: Aust. Nat. Univ. Press.

Lowry, D. C. 1970. Geology of the Western Australian part of the Eucla Basin. *Bull. geol. Surv. West. Aust.* 122:1–200.

Pianka, E. R. 1969a. Sympatry of desert lizards (*Ctenotus*) in Western Australia. *Ecology* 50:1012–1013.

———. 1969b. Habitat specificity, speciation and species density in Australian desert lizards. *Ecology* 50:498–502.

Raven, P. H. 1972. An introduction to continental drift. *Aust. nat. Hist.* 17:245–248.

Raven, P. H., and Axelrod, D. I. 1972. Plate tectonics and Australasian palaeobiogeography. *Science, N.Y.* 176:1379–1386.

Skottsberg, C. 1956. *The natural history of Juan Fernández and Easter Island. I(ii) Derivation of the flora and fauna of Easter Island.* Uppsala: Almqvist & Wiksell.

4. Evolution of Arid Vegetation in Tropical America

Guillermo Sarmiento

Introduction

More or less continuous arid regions cover extensive areas in the middle latitudes of both South and North America, forming a complex pattern of subtropical, temperate, and cold deserts on the western side of the two American continents. They appear somewhat intermingled with wetter ecosystems wherever more favorable habitats occur. These two arid zones are widely separated from each other, leaving a huge gap extending over almost the whole intertropical region (see fig. 4-1). South American arid zones, however, penetrate deeply into intertropical latitudes from northwestern Argentina through Chile, Bolivia, and Peru to southern Ecuador. But they occur either as high-altitude deserts, such as the Puna (high Andean plateaus over 3,000 m), or as coastal fog deserts, such as the Atacama Desert in Chile and Peru, the driest American area. This coastal region, in spite of its latitudinal position and low elevation, cannot be considered as a tropical warm desert, because its cool maritime climate is determined by almost permanent fog. In fact, in most of tropical America, either in the lowlands or in the high mountain chains, from southern Ecuador to southern Mexico, more humid climates and ecosystems prevail. In sharp contrast with the range areas of western North America and the high cordilleras and plateaus of western South America, the tropical American mountains lie in regions of wet climates from their piedmonts to the highest summits. The same is true for the lower ranges located in the interior of the Guianan and Brazilian plateaus.

Upon closer examination, however, it is apparent that, although warm, tropical rain forests and mountain forests, as well as savannas, are characteristic of most of the tropical American landscape, the arid

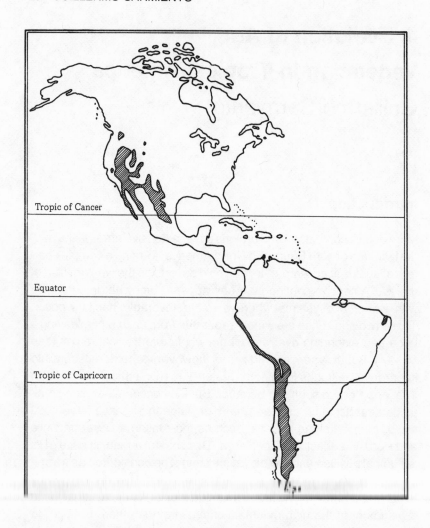

Tropic of Cancer

Equator

Tropic of Capricorn

Fig. 4-1. American arid lands (after Meigs, 1953, modified).

ecosystems are far from being completely absent. If we look at a generalized map of arid-land distribution, such as that of Meigs (1953), we will notice two arid zones in tropical South America: one in northeastern Brazil and the other forming a narrow belt along the Caribbean coast of northern South America, including various small nearby islands. These two tropical areas share some common geographical features:

1. They are quite isolated from each other and from the two principal desert areas in North and South America. The actual distance between the northeast Brazilian arid Caatinga and the nearest desert in the Andean plateaus is about 2,500 km, while its distance from the Caribbean arid region is over 3,000 km. The distance from the Caribbean arid zone to the nearest South American continuous desert, in southern Ecuador, and to the closest North American continuous desert, in central Mexico, is in both cases around 1,700 km.

2. They appear completely encircled by tropical wet climates and plant formations.

3. The two areas are more or less disconnected from the spinal cord of the continent (the Andes cordillera), particularly in the case of the Brazilian region. This fact surely has had major biogeographical consequences.

Recently, interest in ecological research in American arid regions has been renewed, mainly through the wide scope and interdisciplinary research programs of the International Biological Program (Lowe et al., 1973). These studies give strong emphasis to a thorough comparison of temperate deserts in the middle latitudes of North and South America, with the purposes of disclosing the precise nature of their ecological and biogeographical relationships and also of assessing the degree of evolutionary convergence and divergence between corresponding ecosystems and between species of similar ecological behavior. Within this context, a deeper knowledge of tropical American arid ecosystems would provide additional valuable information to clarify some of the previous points, besides having a research interest per se, as a particular case of evolution of arid and semiarid ecosystems of Neotropical origin under the peculiar environmental conditions of the lowland tropics.

The aim of this paper is to present certain available data concerning tropical American arid and semiarid ecosystems, with particular reference to their flora, environment, and vegetation structure. The geographical scope will be restricted to the Caribbean dry region, of which I have direct field knowledge; there will be only occasional further reference to the Brazilian dry vegetation. The Caribbean dry region is still scarcely known outside the countries involved; a review book on arid lands, such as that of McGinnies, Goldman, and Paylore (1968), does not provide a single datum about this region.

In order to delimit more precisely the region I am talking about, a climatic and a vegetational criterion will be used. My field experience suggests that most dry ecosystems in this part of the world lie inside the 800-mm annual rainfall line, with the most arid types occurring below the 500-mm rainfall line. Figure 4-2 shows the course of these two climatic lines through the Caribbean area. Though some wetter ecosystems are included within this limit, particularly at high altitudes, few arid types appear outside this area except localized edaphic types on saline soils, beaches, coral reefs, dunes, or rock outcrops. Only in the Lesser Antilles does a coastal arid vegetation appear under higher rainfall figures, up to 1,200 mm, and this only on very permeable and dry soils near the sea (Stehlé, 1945).

This climatically dry region extends over northern Colombia and Venezuela and covers most of the small islands of the Netherlands Antilles—Aruba, Curaçao, and Bonaire—reaching a total area of about 50,000 km². The nearest isolated dry region toward the northwest is in Guatemala, 1,600 km away; in the north, a dry region is in Jamaica and Hispaniola, 800 km across the Caribbean Sea; while southward the nearest dry region is in Ecuador, 1,700 km away.

From the point of view of vegetation, only the extremes of the Seasonal Evergreen Formation Series and the Dry Evergreen Formation Series of Beard (1944, 1955) will be considered here, including the following four formations: Thorn Woodland, Thorn Scrub, Desert, and Dry Evergreen Bushland. Several papers have dealt with the vegetation of this dry area, but they analyze either only a restricted zone inside this whole region, as those of Dugand (1941, 1970), Tamayo (1941), Marcuzzi (1956), Stoffers (1956), and several others, or they are generalized accounts of plant cover for a whole country that include a short description of the arid types, like those of Cuatrecasas (1958) or Pittier (1926). The aim of this paper is to go one step further than previous investigations—first, considering the entire

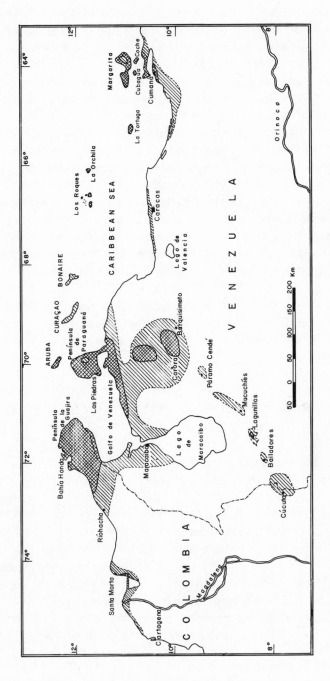

Fig. 4-2. Caribbean arid lands. Semiarid (500–800 mm rainfall) and arid zones (less than 500 mm rainfall) have been distinguished.

Caribbean dry region and, second, comparing it to the rest of American dry lands. My previous paper (1972) had a similar approach. The thorn forests and thorn scrub of tropical America were included in a floristic and ecological comparison between tropical and subtropical seasonal plant formations of South America. I will follow that approach here, but will restrict my scope to the dry extreme of the tropical American vegetation gradient.

To avoid a possible terminological misunderstanding as to certain concepts I am employing, it is necessary to point out that the words *arid* and *semiarid* refer to climatological concepts and will be applied both to climates and to plant formations and ecosystems occurring under these climates. *Dry* will refer to every type of xeromorphic vegetation, either climatically or edaphically determined, such as those of sand beaches, dunes, and rock pavements. *Desert* will be used in its wide geographical sense, that is, a region of dry climate where several types of dry plant formations occur, among them semidesert and desert formations. In this way, for instance, the Sonora and the Monte deserts have mainly a semidesert vegetation, while the Chile-Peru coastal desert shows mainly a desert plant formation. Each time I refer to a *desert vegetation* in contrast to a *desert region*, I shall clarify the point.

The Environment in the Caribbean Dry Lands

Geography

The Caribbean dry region, as its name suggests, is closely linked with the Caribbean coast of northern South America, stretching almost continuously from the Araya Peninsula in Venezuela, at long. 64° W, to a few kilometers north of Cartagena in Colombia, at long. 75° W. Along most of this coast the dry zone constitutes only a narrow fringe between the sea and the forest formation beginning on the lower slopes of the contiguous mountains: the Caribbean or Coast Range in Venezuela and the Sierra Nevada of Santa Marta in Colombia. In many places this arid fringe is no more than a few hundred meters wide. But in the two northernmost outgrowths of the South American continent, the Guajira and Paraguaná peninsulas, the dry region widens to cover these two territories almost completely (see fig. 4-2).

Besides these strictly coastal areas, dry vegetation penetrates deeper inside the hinterland around the northern part of the Maracaibo basin as well as in the neighboring region of low mountains and inner depressions known as the Lara-Falcón dry area of Venezuela. In this zone the aridity reaches more than 200 km from the coast.

Besides this almost continuous dry area in continental South America, the Caribbean dry region extends over the nearby islands along the Venezuelan coast, from Aruba through Curaçao, Bonaire, Los Roques, La Orchila, and other minor islands to Margarita, Cubagua, and Coche. The islands farthest from the continental coast lie 140 km off the Venezuelan coast. Dry vegetation entirely covers these islands, except for a few summits with an altitude over 500 m. The Lesser Antilles somehow connect this dry area with the dry regions of Hispaniola, Cuba, and Jamaica, because almost all of them show restricted zones of dry vegetation (Stehlé, 1945).

Both on the continents and in the islands dry plant formations occupy the lowlands, ranging in altitude from sea level to no more than 600–700 m, covering in this low climatic belt all sorts of land forms, rock substrata, and geomorphological units, such as coastal plains, alluvial and lacustrine plains, early and middle Quaternary terraces, rocky slopes, and broken hilly country of different ages. In the islands dry vegetation also occurs on coral reefs, banks, and on the less-extended occurrences of loose volcanic materials.

Apart from the nearly continuous coastal region and its southward extensions, I should point out that a whole series of small patches or "islands" of dry vegetation and climate occurs along the Andes from western Venezuela across Colombia and Ecuador to Peru. These small and isolated arid patches may be divided into two ecologically divergent types according to their thermal climate determined by altitude: those occurring below 1,500–1,800 m that have a warm or megathermal climate and those appearing above that altitude and belonging then to the meso- or microthermal climatic belts. The latter, such as the small dry islands in the Páramo Cendé and the upper Chama and upper Mocoties valleys of the Venezuelan Andes, even though they have low rainfall, have a less-unfavorable water budget because of their comparatively constant low temperature. Therefore, their vegetation has few features in common with the remaining dry Caribbean areas. On the other hand, the lower-altitude dry patches, like the middle Chama valley, the Tachira-Pamplonita depression, and the lower Chicamocha valley, are quite similar to the dry coastal regions in ecology, flora, and vegetation and will be considered in this study as part of the Caribbean dry lands. I shall point out further the biogeographical significance of this archipelago of Andean dry islands connecting the Caribbean dry region with the southern South American deserts.

Throughout the dry area of northern South America, dry plant formations appear bordered by one or other of three different types of vegetation units: tropical drought-deciduous forest, dry evergreen woodland, or littoral formations (mangroves, littoral woodlands, etc.). In the lower Magdalena valley, as well as in certain other partially flooded areas, marshes and other hydrophytic formations are also common, intermingled with thorn woodland or thorn scrub.

Climate

I propose to analyze the prevailing climatic features of the region enclosed within the 800-mm rainfall line, with particular reference to the main climatic factor affecting plant life, that is, the amount of rainfall and its seasonal distribution, but without disregarding other climatic elements that sharply differentiate tropical and extratropical climates, like minimal temperatures, annual cycle of insolation, and thermo- and photoperiodicity. Lahey (1958) provided a detailed discussion about the causes of the dry climates around the Caribbean Sea, and I shall refer to that paper for pertinent meteorological and climatological considerations on this topic. Porras, Andressen, and Pérez (1966) published a detailed study of the climate of the islands of Margarita, Cubagua, and Coche, some of the driest areas of the Caribbean; some of the climatic data I will discuss have been taken from that paper.

As pointed out before, a major part of the region with annual rainfall figures below 800 mm is located in the megathermal belt, below 600–700 m, and has an annual mean temperature above 24°C. A few small patches along the Andes reach higher elevations, up to 1,500–1,800 m, and their annual mean temperatures go down to 20°C, fitting within what has been considered as the mesothermal belt. However, this temperature difference does not seem to introduce significant changes in vegetation physiognomy or ecology.

Mean annual temperatures in coastal and lowland localities range from a regional maximum of 28.7°C in Las Piedras, at sea level, to 24.2°C in Barquisimeto, a hinterland locality at 566-m elevation. Mean temperatures show very slight month-to-month variation (1° to 3.5°C), as is typical for low latitudes. The annual range of extreme temperatures in this ever-warm region is not so wide as in subtropical or temperate dry regions. The recorded absolute regional maximum does not reach 40°C, while the absolute minima are everywhere above 17°C. As we can see, then, in sharp contrast with the case in extratropical conditions, in the dry Caribbean region low temperatures never constitute an ecological limitation to plant life and natural vegetation.

I have already pointed out that, using natural vegetation as a guideline for our definition of aridity, an annual rainfall of 800 mm roughly separates semiarid and arid from humid regions in this part of the world. Excluding edaphically determined vegetation, the most open and sparse vegetation types appear where rainfall figures do not reach 500 mm. The lowest rainfall in the whole area has been recorded in the northern Guajira Peninsula (Bahía Honda: 183 mm) and in the island of La Orchila, which has the absolute minimum rainfall for the region, 150 mm. Rainfall figures below 300 mm also characterize the small islands of Coche and Cubagua and the central and driest part of Margarita. As we can see, these figures are really very low, fully comparable to many desert localities in temperate South and North America, but in our case these rainfall totals occur under constantly high temperatures and, therefore, represent a less-favorable water balance and a greater drought stress upon plant and animal life.

Concerning rainfall patterns, figure 4-3 shows the rainfall regime at eight localities, arranged in an east-to-west sequence from Cumaná at long. 64°11′ W to Pueblo Viejo at long. 74° 16′ W. The rainfall pattern varies somewhat among the localities appearing in the figure; some places show a unimodal distribution, with the yearly maximum slightly preceding the winter solstice (October to December), while other localities show a bimodal distribution, with a secondary maximum during the high sun period (May to June). It is clear, nevertheless, that all localities have a continuous drought throughout the year, with ten to twelve successive months when rainfall does not reach 100 mm and five to eight months with monthly rainfall figures below 50

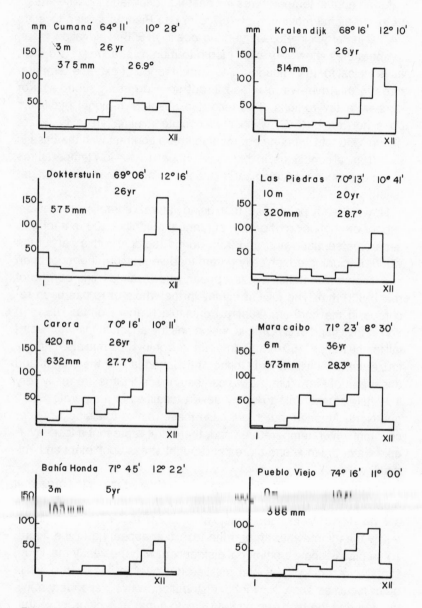

Fig. 4-3. Rainfall regimen for eight stations in the dry Caribbean region. Each climadiagram shows longitude, latitude, altitude, years of recording, mean annual rainfall, and mean temperature.

mm. The total number of rain days ranges over the whole area from forty to sixty.

As is typical for dry climates, rainfall variability is very high, reaching values of 40 percent and more where rainfall is less than 500 mm. This high interannual variability maintains the dryness of the climate and the drought stress upon perennial organisms.

In contrast to most other dry regions in temperate America, relative humidity in the Caribbean dry region is not as low, showing average monthly figures of 70 to 80 percent throughout the year and minimal monthly values of around 55 to 60 percent. Annual pan evaporation, however, is very high, generally exceeding 2,000 mm, with many areas having values as high as 2,500–2,800 mm. Potential evapotranspiration, calculated according to the Thornthwaite formula, reaches 1,600–1,800 mm.

According to its rainfall and temperature, this Caribbean dry region falls within the BSh and BWh climatic types of Koeppen's classification, that is, *hot steppe* and *hot desert* climates. We should remember that in this tropical region the rainfall value setting apart dry and humid climates in the Koeppen system is around 800 mm, which is precisely the limit I have taken according to natural vegetation. In turn, the 400-to-450-mm rainfall line separates BS, or semiarid, from BW, or arid, climates. Following the second system of climatic classification of Thornthwaite, this region comes within the DA' and EA' types, that is, *semiarid megathermal* and *arid megathermal* climates respectively. It is interesting that in both systems the climates corresponding to the dry Caribbean area are the same as those found in the dry subtropical regions of America, such as the Chaco and Monte regions of South America and the Sonoran and Chihuahuan deserts of North America.

We will now consider certain rhythmic environmental factors which influence both climate and ecological behavior of organisms, such as incoming solar radiation and length of day. The sharp contrast between incoming solar radiation at sea level (maximal theoretical values disregarding cloudiness) in low and middle latitudes is well known. At low latitudes daily insolation varies only slightly throughout the year, forming a bimodal curve in correspondence with the sun passing twice a year over that latitude. The total variation between the extremes of maximal and minimal solar radiation during the year is in the order of 50 percent. At middle latitudes the annual radiation curve is unimodal in shape and shows, at a latitude as low as 30° north where North American warm deserts are more widespread, a seasonal variation between extremes in the order of 300 percent.

Photoperiodicity is also inconspicuous in tropical latitudes. At 10° north or south the difference between the shortest and the longest day of the year is around one hour, while at 30° it is almost four hours. Summarizing the climatic data, we can see that the Caribbean dry region has semiarid and arid climates partly comparable to those found in middle-latitude American deserts, particularly insofar as permanent water deficiency is concerned; but these tropical climates differ from the subtropical dry climates by more uniform distribution of solar radiation, higher relative humidity, higher minimal temperatures, slight variation of monthly means, and shorter variation in the length of day throughout the year.

Physiognomy and Structure of Plant Formations

General

Beard (1955) has given the most valuable and widely used of the classifications of tropical American vegetation. Arid vegetation appears as the dry extreme of two series of plant formations: the Seasonal Evergreen Formation Series and the Dry Evergreen Formation Series. Seasonal formations were arranged in Beard's scheme along a gradient of increasing climatic seasonality (rainfall seasonality because all are isothermal climates), from an over-wet regime without dry seasons to the most highly desert types. The successive terms of this series, beginning with the Tropical Rain Forest *sensu stricto* as the optimal plant formation in tropical America are Seasonal Evergreen Forest, Semideciduous Forest, Deciduous Forest, Thorn Woodland, Thorn or Cactus Scrub, and Desert. The first two units appear under slightly seasonal climates; Deciduous Forest together with savannas appear under tropical wet and dry climates; while the last three members of this series, Thorn Woodland, Thorn Scrub, and Desert, occur under dry climates with an extended rainless season and as such are common in the dry Caribbean area.

The Dry Evergreen Formation Series of Beard's classification, in contrast with the Seasonal Series, occurs under almost continuously dry climates but where monthly rainfall values are not as low as during the dry season of the seasonal climates. The driest formations on this series are the Thorn Scrub and the Desert formations, these two series being convergent in physiognomic and morphoecological features according to Beard and other authors, such as Loveless and Asprey (1957). The remaining less-dry type, next to the previous two, is the Dry Evergreen Bushland formation, which also occurs under the dry climates of the Caribbean area. In summary, dry vegetation in tropical America has been included in four plant formations: Dry Evergreen Bushland, Thorn Woodland, Thorn or Cactus Scrub, and Desert. Their structures according to the original definitions are represented in figure 4-4. All of them have open physiognomies, where the upper-layer canopy in the more structured and richer types does not surpass 10 m in height and a cover of 80 percent, decreasing then in height and cover as the environmental conditions become less favorable.

Plant formations occurring in the arid Caribbean region fit closely with Beard's classification and types, though it seems necessary to add a new formation: Deciduous Bushland, structurally equivalent to the Dry Evergreen Bushland, but with a predominance of deciduous woody species. Before going into some details about each dry plant formation in the Caribbean area, let me add a final remark about the evident difficulty met with when some of the vegetation is classified in one or another type, particularly in the case of some low and poor associations of the Tropical Deciduous Forest whose features overlap with those of the Dry Evergreen Bushland or the Thorn Woodland. Human interference, through wood cutting and heavy goat grazing, frequently makes subjective conclusions difficult, and in many instances only a thorough quantitative recording of vegetational features could allow an objective characterization and classification of the stand. At a preliminary survey level these doubts remain. A detailed study of dry plant formations, such as that of Loveless and Asprey (1957, 1958) in Jamaica, will emphasize the need for quantitative data on vegetation structure and species morphoecology in order to classify these difficult intermediate dry formations.

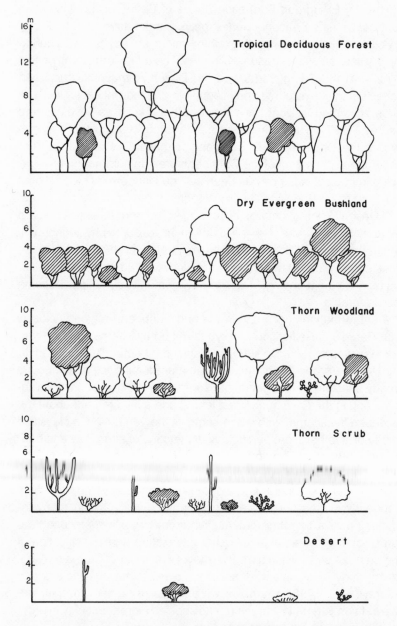

Fig. 4-4. Vegetation profiles of tropical American dry formations. Tropical Deciduous Forest has been included for comparison.

I will now very briefly consider each of the five dry plant formations as they occur nowadays in the Caribbean dry region.

Dry Evergreen Bushland

The Dry Evergreen Bushland formation has a closed canopy of low trees and shrubs at a height of about 2 to 4 m. Sparse taller trees and cacti, up to 10 m high, may emerge from this canopy. The two essential physiognomic features of this plant formation are, first, the closed nature of the plant cover, leaving no bare ground at all, and, second, the predominance of evergreen species, with a minor proportion of deciduous and succulent-aphyllous elements. The dominant woody species are evergreen low trees and shrubs, with sclerophyllous medium-sized leaves.

Floristically it is a rather rich plant formation, taking into account its dry nature, with an evident differentiation into various floristic associations. The most important families of this formation are the Euphorbiaceae, Boraginaceae, Capparidaceae, Leguminosae (Papilionoideae and Caesalpinoideae), Rhamnaceae, Polygonaceae, Rubiaceae, Myrtaceae, Flacourtiaceae, and Celastraceae. Cacti and agaves are also frequent, interspersed with a rich subshrubby and herbaceous flora.

This formation is widespread in the whole Caribbean dry region, being frequent in the islands, in the mainland coasts, and in the small Andean arid patches. Its physiognomy clearly differentiates this evergreen bushland from all other tropical dry formations; and from this physiognomic and structural viewpoint it looks more like the temperate scrubs in Mediterranean climates, such as the low chaparral of California and the garigue of southern France, than most other tropical types.

Deciduous Bushland

The Deciduous Bushland is quite similar in structure to the Dry Evergreen Bushland, but it differs mainly by the predominance of deciduous shrubs, while evergreen and aphyllous species only share a secondary role. This gives a highly seasonal appearance to the Deciduous Bushland, with two acutely contrasting aspects: one during the leafless period and the other when the dominant species are in full leaf. It also differs from other seasonal dry formations, such as the Thorn Woodland and the Thorn Scrub, because of its closed canopy of shrubs and low trees that leaves no bare ground. The most common families in this formation are the Leguminosae (mainly Mimosoideae), Verbenaceae, Euphorbiaceae, and Cactaceae. Floristically this type is not well known, but apparently it differs sharply from the Thorn Woodland and Thorn Scrub. Up to date the Deciduous Bushland has only been reported in the Lara-Falcón area (Smith, 1972).

Thorn Woodland

The distinctive physiognomic feature of the Thorn Woodland is a lack of a continuous canopy at any height, leaving large spaces of bare soil between the sparse trees and shrubs, particularly during dry periods when herbaceous annual cover is lacking. The upper layer of high shrubs and low trees and succulents is from 4 to 8 m high, with a variable cover, from less than 10 percent to a maximum of around 75 percent. A second woody layer 2 to 4 m high is generally the most important in cover, showing values ranging from 30 to 70 percent. The shrub layer of 0.5 to 2 m is also conspicuous, inversely related in importance to the two uppermost layers. The total cover of the herb and soil layers varies during the year because of the seasonal development of annual herbs, geophytes, and hemicryptophytes; the permanent biomass in these lowest layers is given by small cacti, like *Mammillaria*, *Melocactus*, and *Opuntia*.

As for the morphoecological features of its species, this formation is characterized by a high proportion of thorny elements, by many succulent shrubs, and by a total dominance of the smallest leaf sizes (lepto- and nanophyll), with a smaller proportion of aphyllous and microphyllous species together with rare mesophyllous elements; the last mentioned are generally highly scleromorphic. The relative proportion of evergreen and deciduous species is almost the same, with a good proportion of brevideciduous species.

From the floristic aspect this formation has a very characteristic flora, scarcely represented in wetter plant types. Among the most important families are the Leguminosae (particularly Mimosoideae and

Caesalpinoideae), Cactaceae, Capparidaceae, and Euphorbiaceae. Many floristic associations can be distinguished on the basis of the dominant species, but their distribution and ecology are scarcely known. The most important single species in this formation, distributed over its area, is undoubtedly *Prosopis juliflora*. When it is present, this low tree usually shares a dominant role in the community. This may probably be due, among other reasons, to its noteworthy ability for regrowth after cutting, as well as to its unpalatability to all domestic herbivores.

Thorn Scrub or Cactus Scrub

The Thorn Scrub, equivalent to the Semidesert formation of arid temperate areas, is still lower and more sparse than the Thorn Woodland, leaving a major part of bare ground, particularly during the driest period of the year. Low trees and columnar cacti from 4 to 8 m high appear widely dispersed or are completely lacking. Shrubs from 0.5 to 2 m high, though they form the closest plant layer, are also widely separated, as well as the subshrubs and herbs that form the scattered lower layer. Floristically the Thorn Scrub seems to be an impoverished Thorn Woodland, without significant additions to the flora of that formation. Cactaceae, Capparidaceae, Euphorbiaceae, and Mimosoideae continue to be the best-represented taxa. Even by its morphoecology and functionality this formation resembles the Thorn Woodland, showing a heterogeneous mixture of evergreen, deciduous, brevideciduous, and aphyllous species, with the smallest leaf sizes frequently being of sclerophyllous texture. Succulent species, particularly cacti, appear here at their optimum, frequently being the most noteworthy feature in the physiognomy of the plant formation.

This Thorn Scrub physiognomy is not so widely found in the Caribbean arid region as in the temperate deserts of North and South America. By structure and biomass it is comparable to the Semidesert formations of those arid regions, though the most extended associations of temperate American deserts, those formed by nonspiny low shrubs such as *Larrea*, are completely absent from the tropical American area. Thorn Scrub occurs in the Lara-Falcón region of Venezuela, in the northernmost part of the Guajira Peninsula, and in the driest islands like Coche and Cubagua.

Desert

Extremely desertic vegetational physiognomies are not uncommon in the Caribbean arid zone, but most of them seem determined by substratum-related factors and not primarily by climate. Thus, for example, one of the most widespread types of Desert formation occurs in the Lara-Falcón area, on sandstone hills of Tertiary age. Only four or five species of low shrubs grow there, such as species of *Cassia*, *Sida*, and *Heliotropium*, very widely interspersed with some woody *Capparis* and various Cactaceae and Mimosoideae. The total ground cover is less than 2 or 3 percent. To explain this extremely desertic vegetation in an area with enough rainfall to maintain thorn woodland in neighboring situations, Smith (1972) suggested the existence of heavy metals in the rock substrata; but there is not yet any further evidence to sustain this hypothesis, though undoubtedly the responsible factor is linked to a particular type of geological formation.

Another type of desert community that covers a wide extent of flat country in northern Venezuela appears on heavy soil developed on old Quaternary terraces. This desert community scarcely covers more than 5 percent of the ground and is composed mainly of species of *Jatropha*, *Opuntia*, *Lemaireocereus*, and *Ipomoea*, together with some annual herbs. Though this community is rather common in several parts of the Caribbean arid area, a satisfactory explanation for its occurrence has not been given for it, either.

Some more easily understood types of Desert formation are the salt deserts near the coast and the sand deserts of dunes and beaches. Salt deserts are almost everywhere characterized by low, shrubby Chenopodiaceae, such as *Salicornia* and *Heterostachys*, while sand deserts show a dominance of geophytes together with some shrubby species of *Lycium*, *Castela*, *Opuntia*, and *Acacia*.

Floristic Composition and Diversity

The floristic inventory of the Caribbean arid vegetation has not yet been made. My list of plant families and genera has been compiled from several sources (Boldingh, 1914; Tamayo, 1941 and 1967; Dugand, 1941 and 1970; Pittier et al., 1947; Croizat, 1954; Marcuzzi, 1956; Stoffers, 1956; Cuatrecasas, 1958; Trujillo, 1966) as well as from direct field knowledge of this vegetation and flora.

Table 4-1 presents a list of 94 families and 470 genera which have been reported from this area. Both figures must be taken as rough approximations of the regional total flora, because this arid flora is still not well known and in many areas plant collections are lacking; there are also some overrepresentations in the tabulated figures, because many of the listed taxa collected in the arid region surely belong to various riparian forests and therefore are not strictly part of the arid Caribbean flora. The total number of species is still more imprecisely known; a figure of 1,000 will give an idea of the magnitude of the species diversity in this vegetation.

If the floristic richness and diversity in more restricted areas is taken into consideration, the following figures are obtained: a thorough

---

Table 4-1. *Families and Genera of Flowering Plants Reported from the Caribbean Dry Region*

Family	Genera
Acanthaceae	*Anisacanthus*, *Anthacanthus*, *Dicliptera*, *Elytraria*, *Justicia*, *Odontonema*, *Ruellia*, *Stenandrium*
Achatocarpaceae	*Achatocarpus*
Aizoaceae	*Mollugo*, *Sesuvium*, *Trianthema*
Amaranthaceae	*Achyranthes*, *Alternanthera*, *Amaranthus*, *Celosia*, *Cyathula*, *Froelichia*, *Gomphrena*, *Iresine*, *Pfaffia*, *Philoxerus*
Amaryllidaceae	*Agave*, *Crinum*, *Fourcroya*, *Hippeastrum*, *Hymenocallis*, *Hypoxis*, *Zephyranthes*
Anacardiaceae	*Astronium*, *Mauria*, *Metopium*, *Spondias*
Apocynaceae	*Aspidosperma*, *Echites*, *Forsteronia*, *Plumeria*, *Prestonia*, *Rauvolfia*, *Stemmadenia*, *Thevetia*
Araceae	*Philodendron*
Aristolochiaceae	*Aristolochia*

Asclepiadaceae	*Asclepias, Calotropis, Cynanchum, Gomphocarpus, Gonolobus, Ibatia, Marsdenia, Metastelma, Omphalophthalmum, Sarcostemma*
Bignoniaceae	*Amphilophium, Anemopaegma, Arrabidaea, Bignonia, Clytostoma, Crescentia, Distictis, Lundia, Memora, Pithecoctenium, Tabebuia, Tecoma, Xylophragma*
Bombacaceae	*Bombacopsis, Bombax, Cavanillesia, Pseudobombax*
Boraginaceae	*Cordia, Heliotropium, Rochefortia, Tournefortia*
Bromeliaceae	*Aechmea, Bromelia, Pitcairnia, Tillandsia, Vriesia*
Burseraceae	*Bursera, Protium*
Cactaceae	*Acanthocereus, Cephalocereus, Cereus, Hylocereus, Lemaireocereus, Mammillaria, Melocactus, Opuntia, Pereskia, Phyllocactus, Rhipsalis*
Canellaceae	*Canella*
Capparidaceae	*Belencita, Capparis, Cleome, Crataeva, Morisonia, Steriphoma, Stuebelia*
Caryophyllaceae	*Drymaria*
Celastraceae	*Hippocratea, Maytenus, Pristimera, Rhacoma, Schaefferia*
Chenopodiaceae	*Atriplex, Chenopodium, Heterostachys, Salicornia*
Cochlospermaceae	*Amoreuxia, Cochlospermum*
Combretaceae	*Bucida, Combretum*
Commelinaceae	*Callisia, Commelina, Tripogandra*
Compositae	*Acanthospermum, Ambrosia, Aster, Baltimora, Bidens, Conyza, Egletes, Eleutheranthera, Elvira, Eupatorium, Flaveria,*

*Gundlachia, Isocarpha, Lactuca, Lagascea,
Lepidesmia, Lycoseris, Mikania, Oxycarpha,
Parthenium, Pectis, Pollalesta, Porophyllum,
Sclerocarpus, Simsia, Sonchus, Spilanthes,
Synedrella, Tagetes, Trixis, Verbesina,
Vernonia, Wedelia*

Convolvulaceae	*Bonomia, Cuscuta, Evolvulus, Ipomoea, Jacquemontia, Merremia*
Cruciferae	*Greggia*
Cucurbitaceae	*Bryonia, Ceratosanthes, Corallocarpus, Doyerea, Luffa, Melothria, Momordica, Rytidostylis*
Cyperaceae	*Bulbostylis, Cyperus, Eleocharis, Fimbristylis, Hemicarpha, Scleria*
Elaeocarpaceae	*Muntingia*
Erythroxylaceae	*Erythroxylon*
Euphorbiaceae	*Acalypha, Actinostemon, Adelia, Argithamnia, Bernardia, Chamaesyce, Cnidoscolus, Croton, Dalechampsia, Ditaxis, Euphorbia, Hippomane, Jatropha, Julocroton, Mabea, Manihot, Pedilanthus, Phyllanthus, Sebastiania, Tragia*
Flacourtiaceae	*Casearia, Hecatostemon, Laetia, Mayna*
Gentianaceae	*Enicostemma*
Gesneriaceae	*Kohleria, Rechsteineria*
Goodeniaceae	*Scaevola*
Gramineae	*Andropogon, Anthephora, Aristida, Bouteloua, Cenchrus, Chloris, Cynodon, Dactyloctenium, Digitaria, Echinochloa, Eleusine, Eragrostis, Eriochloa, Leptochloa, Leptothrium, Panicum, Pappophorum, Paspalum, Setaria, Sporobolus, Tragus, Trichloris*
Guttiferae	*Clusia*

Hernandaceae	*Gyrocarpus*
Hydrophyllaceae	*Hydrolea*
Krameriaceae	*Krameria*
Labiatae	*Eriope, Hyptis, Leonotis, Marsypianthes, Ocimum, Perilomia, Salvia*
Lecythidaceae	*Chytroma, Lecythis*
Leguminosae (Caesalpinoideae)	*Bauhinia, Brasilettia, Brownea, Caesalpinia, Cassia, Cercidium, Haematoxylon, Schnella*
Leguminosae (Mimosoideae)	*Acacia, Calliandra, Cathormium, Desmanthus, Inga, Leucaena, Mimosa, Piptadenia, Pithecellobium, Prosopis*
Leguminosae (Papilionoideae)	*Abrus, Aeschynomene, Benthamantha, Callistylon, Canavalia, Centrosema, Crotalaria, Dalbergia, Dalea, Desmodium, Diphysa, Erythrina, Galactia, Geoffraea, Gliricidia, Humboldtiella, Indigofera, Lonchocarpus, Machaerium, Margaritolobium, Myrospermum, Peltophorum, Phaseolus, Piscidia, Platymiscium, Pterocarpus, Rhynchosia, Sesbania, Sophora, Stizolobium, Stylosanthes, Tephrosia*
Lennoaceae	*Lennoa*
Liliaceae	*Smilax, Yucca*
Loasaceae	*Mentzelia*
Loganiaceae	*Spigelia*
Loranthaceae	*Oryctanthus, Phoradendron, Phthirusa, Struthanthus*
Lythraceae	*Ammannia, Cuphea, Pleurophora, Rotala*
Malpighiaceae	*Banisteria, Banisteriopsis, Brachypteris, Bunchosia, Byrsonima, Heteropteris, Hiraea, Malpighia, Mascagnia, Stigmatophyllum, Tetrapteris*

Malvaceae	*Abutilon, Bastardia, Cienfuegosia, Hibiscus, Malachra, Malvastrum, Pavonia, Sida, Thespesia, Urena, Wissadula*
Melastomaceae	*Miconia, Tibouchina*
Meliaceae	*Trichilia*
Menispermaceae	*Cissampelos*
Moraceae	*Brosimum, Chlorophora, Ficus, Helicostylis*
Myrtaceae	*Anamomis, Pimenta, Psidium*
Nyctaginaceae	*Allionia, Boerhavia, Mirabilis, Naea, Pisonia, Torrubia*
Ochnaceae	*Sauvagesia*
Oenotheraceae	*Jussiaea*
Olacaceae	*Schoepfia, Ximenia*
Oleaceae	*Forestiera, Linociera*
Opiliaceae	*Agonandra*
Orchidaceae	*Bifrenaria, Bletia, Brassavola, Brassia, Catasetum, Dichaea, Elleanthus, Epidendrum, Gongora, Habenaria, Ionopsis, Maxillaria, Oncidium, Pleurothallis, Polystachya, Schombergkia, Spiranthes, Vanilla*
Oxalidaceae	*Oxalis*
Palmae	*Bactris, Copernicia*
Papaveraceae	*Argemone*
Passifloraceae	*Passiflora*
Phytolaccaceae	*Petiveria, Rivinia, Seguieria*
Piperaceae	*Peperomia, Piper*
Plumbaginaceae	*Plumbago*
Polygalaceae	*Bredemeyera, Monnina, Polygala, Securidaca*
Polygonaceae	*Coccoloba, Ruprechtia, Triplaris*

Portulacaceae	*Portulaca, Talinum*
Ranunculaceae	*Clematis*
Rhamnaceae	*Colubrina, Condalia, Gouania, Krugiodendron, Zizyphus*
Rubiaceae	*Antirrhoea, Borreria, Cephalis, Chiococca, Coutarea, Diodia, Erithalis, Ernodea, Guettarda, Hamelia, Machaonia, Mitracarpus, Morinda, Psychotria, Randia, Rondeletia, Sickingia, Spermacoce, Strumpfia*
Rutaceae	*Amyris, Cusparia, Esenbeckia, Fagara, Helietta, Pilocarpus*
Sapindaceae	*Allophylus, Cardiospermum, Dodonaea, Paullinia, Serjania, Talisia, Thinouia, Urvillea*
Sapotaceae	*Bumelia, Dipholis*
Scrophulariaceae	*Capraria, Ilysanthes, Scoparia, Stemodia*
Simarubaceae	*Castela, Suriana*
Solanaceae	*Bassovia, Brachistus, Capsicum, Cestrum, Datura, Lycium, Nicotiana, Physalis, Solanum*
Sterculiaceae	*Ayenia, Buettneria, Guazuma, Helicteres, Melochia, Waltheria*
Theophrastaceae	*Jacquinia*
Tiliaceae	*Corchorus, Triumfetta*
Turneraceae	*Piriqueta, Turnera*
Ulmaceae	*Celtis, Phyllostylon*
Urticaceae	*Fleurya*
Verbenaceae	*Aegiphila, Bouchea, Citharexylon, Clerodendrum, Lantana, Lippia, Phyla, Priva, Stachytarpheta, Vitex*
Violaceae	*Rinorea*
Vitaceae	*Cissus*

| Zingiberaceae | *Costus* |
| Zygophyllaceae | *Bulnesia, Guaiacum, Kallstroemia, Tribulus* |

---

floristic survey of a dry forest community in the lower Magdalena valley in Colombia, with an annual rainfall of 720 mm (Dugand, 1970), gives a total of 55 families, 154 genera, and 187 species of flowering plants in a stand of less than 300 ha. For the three small islands of Curaçao, Aruba, and Bonaire, with a total area of 860 km², Boldingh (1914) gives a list of 79 families, 239 genera, and 391 species of flowering plants, excluding the mangroves as the only local formation not belonging to the dry types.

As we can see in table 4-1, the best-represented families in total number of genera are the Leguminosae (50), Compositae (33), Euphorbiaceae (20), and Rubiaceae (19). Other well-represented families are the Amaranthaceae, Malvaceae, Malpighiaceae, Cactaceae, Verbenaceae, Orchidaceae, and Asclepiadaceae; almost all of them are typical of warm, arid floras everywhere.

If we compare now the floristic richness of this Caribbean dry region to the flora of North and South American middle-latitude deserts, we obtain roughly equivalent figures. In fact, Johnson (1968) gives a total of 278 genera and 1,084 species for the Mojave and Colorado deserts of California; Shreve (1951) reports 416 genera for the whole Sonoran desert, while Morello (1958) gives a list of 160 genera from the floristically less known Monte desert in Argentina. We can see then, that, in spite of a smaller total area, the Caribbean dry flora is as rich as other American desert floras.

Johnson (1968) gives a list of monotypic or ditypic genera of the Mojave-Colorado deserts, considered according to the ideas of Stebbins and Major (1965) to be old relict taxa. That list includes 60 species belonging to 56 genera. Applying this same criterion to the flora of the Caribbean desert I have recognized only 14 relict endemic species— a number that, even if it represents a gross underestimate, is significantly smaller than the preceding one (see table 4-2).

Concerning the geographic distribution and centers of diversification of the Caribbean arid taxa, it is not possible to proceed here to a detailed analysis because of the fragmentary knowledge of plant dis-

tribution in tropical America. However, I have tried to give a preliminary analysis based on only a few best-known families.

Taking the Compositae for instance, one of the most diversified
families within this vegetation, I took the data on its distribution from
Aristeguieta (1964) and Willis (1966). The species of 19 genera occurring in the Caribbean dry lands could be considered as widely distributed weeds, whose areas also extend to arid climates. Six genera
are very rich genera with a few species also occurring in arid vegetation: *Eupatorium*, *Vernonia*, *Mikania*, *Aster*, *Verbesina*, and *Simsia*;
2 genera (*Lepidesmia* and *Oxycarpha*) are monotypic taxa endemic
to the Caribbean coasts; while the remaining 6 genera (*Pollalesta*,
Egletes, *Baltimora*, *Gundlachia*, *Lycoseris*, and *Sclerocarpus*) are
small-to-medium-sized taxa restricted to tropical America, with some
species characteristic of arid plant formations. We see, then, that in
this family an important proportion of the species that occur in the
arid vegetation may be considered as weeds (19 out of 33 genera);
one part (6/33) has originated from widely distributed and very rich
genera, some of whose species have succeeded in colonizing arid
habitats also; while the remaining part, about a quarter of the genera
of Compositae occurring in arid vegetation, is formed of species belonging to genera of more restricted distribution and lesser adaptive
radiation, whose presence in this arid flora may be indicative of the
adaptation to arid conditions of an ancient Neotropical floristic stock—
in some cases, as in the two monotypic endemics, probably through
a rather long evolution in contact with similar environmental stress.

In all events, the Compositae, a very important family in the temperate and cold American deserts, neither shows a similar degree of
differentiation in the arid Caribbean flora nor occupies a prominent
role in these tropical plant communities.

Another family whose taxonomy and geographical distribution is
rather well known, the Bromeliaceae (Smith, 1971), has five genera
inhabiting the Caribbean arid lands; three of them, *Pitcairnia*, *Vriesia*,
and *Aechmea*, are very rich genera (150 to 240 species) mainly growing in humid vegetation types but with a few species also entering dry
plant formations. None of them is exclusive to the arid types. *Tillandsia*, a great and polymorphous genus of more than 350 species
adapted to nearly all habitat types from the epiphytic types in the rain
forests to the xeric terrestrial plants of extreme deserts, has 15

Table 4-2. *Relictual Endemic Species Occurring in the Caribbean Dry Region*

Family	Species
Asclepiadaceae	*Omphalophthalmum ruber* Karst.
Capparidaceae	*Belencita hagenii* Karst.
Capparidaceae	*Stuebelia nemorosa* (Jacq.) Dugand
Compositae	*Lepidesmia squarrosa* Klatt
Compositae	*Oxycarpha suaedaefolia* Blake
Cucurbitaceae	*Anguriopsis (Doyerea) margaritensis* Johnson
Leguminosae	*Callistylon arboreum* (Griseb.) Pittier
Leguminosae	*Humboldtiella arborea* (Griseb.) Hermann
Leguminosae	*Humboldtiella ferruginea* (H.B.K.) Harms.
Leguminosae	*Margaritolobium luteum* (Johnson) Harms.
Leguminosae	*Myrospermum frutescens* Jacq.
Lennoaceae	*Lennoa caerulea* (H.B.K.) Fourn.
Rhamnaceae	*Krugiodendron ferreum* (Vahl.) Urb.
Rubiaceae	*Strumpfia maritima* Jacq.

species recorded in the Caribbean arid lands; 14 of them are widely distributed species also occurring in dry formations. Only 1 species, *T. andreana*, growing on bare rock, seems strictly confined to dry plant formations. The fifth genus, *Bromelia*, a medium-sized genus of about 40 species, has 4 species growing in deciduous forests in the Caribbean that also extend their areas to the drier plant formations. As we can see by the distribution patterns of this old Neotropical family, the degree of speciation that has occurred in response to aridity in the Caribbean region seems to be minimal. This fact is in sharp contrast with the behavior of this family in other South American deserts, such as the Monte and the Chilean-Peruvian coastal deserts, where it has reached a good degree of diversification.

Let us take as a last example a typical family of arid lands, the Zygophyllaceae, recently studied by Lasser (1971) in Venezuela; it has four genera growing in the Caribbean arid region of which two,

Kallstroemia and *Tribulus*, are weedy genera of widely distributed species on bare soils and in dry habitats. The other two genera, *Bulnesia* and *Guaiacum*, are typical elements of arid and semiarid Neotropical plant formations. *Bulnesia* has its maximal diversification in semiarid and arid zones of temperate South America; while only one species, *B. arborea*, has reached the deciduous forests and thorn woodlands of northern South America, but without extending even to the nearby islands. But it is a dominant tree in many thorn woodland communities of northern South America. *Guaiacum* is a peri-Caribbean genus with several species from Florida to Venezuela, some of them exclusively restricted to arid coastal vegetation. In summary, this small family, whose species are frequently restricted to dry regions, does not show in the Caribbean arid flora the same degree of differentiation it has attained in southern South America, but it has nevertheless distinctly arid species, some originating from the south, such as *Bulnesia*, others from the north, such as *Guaiacum*.

Conclusions

As a conclusion, I wish to point out the most significant facts that follow from the preceding data. We have seen that in northern South America and in the nearby Caribbean islands a region of dry climates exists, which includes semiarid and arid climatic types, wherein five different plant formations occur. Considering the major environmental feature acting upon plant and animal life in this area, that is, the strong annual water deficit, these ecosystems seem subjected to water stress of comparable intensity and extension to that influencing living organisms in the subtropical South and North American deserts. If this water stress constitutes the directing selective force in the evolution of plant species and vegetation forms, the evolutionary framework would be comparable in tropical and extratropical American deserts. If, therefore, significant differences in speciation and vegetation features between these ecosystems could be detected, either they ought to be attributed to a different period of evolution under similar selective pressures, in which case the tropical and temperate American deserts would be of noncomparable geological age, or they could be attributed to the action on the evolution of these species of other environmental factors linked to the latitudinal difference between these deserts.

As many floristic and ecological features of these two types of ecosystems do not seem to be quite similar, even at a preliminary qualitative level of comparison, both previous hypotheses, that of differential age and that of divergent environmental selection, could probably be true. This supposes that the ancestral floristic stock feeding all dry American warm ecosystems was not so different as to explain the actual divergences on the basis of this sole historical factor.

The structure and physiognomy of plant formations occurring in the Caribbean area under a severe arid climate do not seem to correspond strictly to most semidesertic or desertic physiognomies of temperate North and South America. Several plant associations show undoubtedly a high degree of physiognomic convergence, also emphasized by a close floristic affinity, as is the case of the thorn scrub communities dominated in all these regions by species of *Prosopis*, *Cercidium*, *Cereus*, and *Opuntia*. But the most widespread plant associations in temperate American deserts, which are the scrub communities where a mixture of evergreen and deciduous shrubs prevail, like the *Larrea divaricata–Franseria dumosa* association of the Sonoran desert or the *Larrea cuneifolia* communities of the Monte desert; or the communities characterized by aphyllous or subaphyllous shrubs or low trees, such as the *Bulnesia retama–Cassia aphylla* communities of South America or the various *Fouquieria* associations in North America, do not have a similar physiognomic counterpart in tropical America.

As I have already noted in a previous paper (1972), even the degree of morphoecological adaptation in tropical American arid species is significantly smaller than that exhibited by the temperate American desert flora. Such plant features as succulence, spines, or aphylly are widely represented in the desert floras of North and South America, but they appear much more restricted quantitatively in the tropical American arid flora where, for instance, only one family of aphyllous plants occurs, the Cactaceae, in contrast to eleven families in the Monte region of Argentina.

Concerning floristic diversity, the dry Caribbean vegetation has a richness comparable to North American warm-desert floras and perhaps a richer flora than the warm deserts of temperate South America. The tropical arid flora is highly heterogeneous in origin and affinities, with the most significant contribution coming from neighboring less-dry formations, particularly the Tropical Deciduous Forest and the Dry Evergreen Woodland, with an important contribution from cosmopolitan or subcosmopolitan weeds, and a variety of floristic elements whose area of greater diversification occurs in northern or southern latitudes.

Among the elements of direct tropical descent reaching the dry formations from the contiguous less-arid types, the species of wide ecological spectrum predominate, whose ecological amplitude extends from subhumid or seasonally wet climates to semiarid and arid plant formations. On the other hand, few of them show a narrow ecological amplitude, appearing thus restricted only to arid plant communities; and in the majority of these cases the species thus restricted occur in particular types of habitats, like sand beaches, dunes, coral reefs, saline soils, and rock outcrops.

There exist in the Caribbean dry flora some species which are old relictual endemic taxa, in the sense considered by Stebbins and Major (1965), but they are neither as numerous as in North American deserts nor characteristic of "normal" habitats or typical communities; they are, rather, typical species of particular edaphic conditions or characteristics of the less-extreme types, such as the deciduous forests and dry evergreen woodlands.

In summary, then, the speciation of the autochthonous tropical taxa has been important in subhumid or semiarid plant formations as well as in restricted dry habitats, but the arid flora has received only a minor contribution from this source.

In spite of the actual occurrence of a chain of arid islands along the Andes connecting the dry areas of Venezuela and Peru, where neighboring patches occur no more than 200 to 300 km apart, southern floristic affinities are not conspicuous among the families analyzed. Further arguments are available to support this lack of connection between Caribbean and southern South American deserts on the basis of the distributions of all genera of Cactaceae (Sarmiento, 1973, unpublished). The representatives of this typical family that live in the Caribbean region show a closer phylogenetic affinity with the Mexican and West Indian cactus flora, a looser relationship with the Brazilian cactus flora, and a much more restricted affinity with the Peruvian and Argentinean cactus flora.

This slight affinity between tropical American and southern South American dry floras, in spite of more direct biogeographical and paleogeographical connections, is a rather difficult fact to explain, particularly if we consider that some species of disjunct area between North and South America, *Larrea divaricata*, for example, originated in South America and later expanded northward (Hunziker et al., 1972). These species have therefore crossed tropical America, but have not remained there.

In contrast to the loose affinity with southern South America, a stronger relationship with the North American arid flora is easily discernible. The most noteworthy cases are those of the genera *Agave*, *Fourcroya*, and *Yucca*, richly diversified in Mexico and southwestern United States, that reach their southern limits in the dry regions of northern South America. There are many other cases of North American genera, characteristic of dry regions, extending southward to Venezuela, Colombia, or less commonly to Ecuador and northern Brazil.

We can thus infer from the above information that the origin and age of the Caribbean arid vegetation certainly seems heterogeneous. Some elements evolved in tropical dry environments; many are almost cosmopolitan; others came from the north; and a few also came from the south. Several migratory waves along different routes probably occurred during a rather long evolutionary history under similar environmental conditions. Though the Central American connection does not actually offer a natural bridge for arid-adapted species, and there is no evidence of the former existence of this type of biogeographical bridge, the northern affinity of many Caribbean desert elements may be more easily explained by resorting to a dry bridge across the Caribbean islands, from Cuba and Hispaniola through the Lesser Antilles to Venezuela, instead of a more hypothetical Central American pass.

Axelrod's model (1950) of gradual evolution of the arid flora and vegetation in southwestern North America from a Madro-Tertiary geoflora, with the most arid forms and the maximal widespread of arid plant formations occurring only during the Quaternary, does not seem to fit well with the evidence provided by the analysis of the arid Caribbean flora and vegetation. On the contrary, the ideas of Stebbins and Major (1965) about the existence of small arid pockets along the western mountains from the late Mesozoic upward, together with a much more agitated evolutionary history from that time on to the Quaternary, are probably in better agreement with these data, which account for a heterogeneous and polychronic origin of these elements.

Acknowledgments

It is a great pleasure for me to acknowledge all the intellectual stimulus, material help, arduous criticism, and audacious ideas received through frequent and passionate discussions of these topics with my colleague, Maximina Monasterio.

Summary

Tropical American arid vegetation, particularly the formations occurring along the Caribbean coast of northern South America and the small nearby islands, is still not well known. However, within the framework of a comparative analysis of all American dry areas, this region provides not only the interest of knowing the features of plant cover in the driest region of tropical America, but also the knowledge that this possibly may clarify many obscure points of Neotropical biogeography, such as the evolutionary history of arid plant formations and the origin of their flora.

The major points of Caribbean dry ecosystems dealt with in this paper are (a) geographical distribution and climatic conditions, mainly the annual water deficiency and some differential features between low- and middle-latitude climates; (b) physiognomy, structure, and morphoecological traits of each of the five plant formations occurring in that area; and (c) floristic richness, origin, and affinities of floristic elements.

On this basis some relevant facts are discussed, such as the lack of correspondence between arid vegetation physiognomies in tropical and temperate American dry regions; the comparable floristic diversification; and the varied origin of its taxa, where most elements evolved on the spot from a tropical drought-adapted stock. Some others are cosmopolitan taxa; many came from North America; and a few came from the south. This brief analysis leads to the hypothesis that tropical American desert flora is, at least in part, of considerable age and shows a heterogeneous origin, probably brought about by several migratory events. All these facts seem to support Stebbins and Major's ideas about the complex evolution of American dry flora and vegetation.

References

Aristeguieta, L. 1964. Compositae. In *Flora de Venezuela, X*, ed. T. Lasser. Caracas: Instituto Botánico.

Axelrod, D. I. 1950. Evolution of desert vegetation in western North America. *Publs Carnegie Instn* 590:1–323.

Beard, J. S. 1944. Climax vegetation in tropical America. *Ecology* 25: 127–158.

———. 1955. The classification of tropical American vegetation-types. *Ecology* 36:89–100.

Boldingh, I. 1914. *The flora of Curaçao, Aruba and Bonaire*, vol. 2. Leiden: E. J. Brill.

Croizat, L. 1954. La faja xerófila del Estado Mérida. *Universitas Emeritensis* 1:100–106.

Cuatrecasas, J. 1958. Aspectos de la vegetación natural de Colombia. *Revta Acad. colomb. Cienc. exact. fís. nat.* 10:221–268.

Dugand, A. 1941. Estudios geobotánicos colombianos. *Revta Acad. colomb. Cienc. exact. fís. nat.* 4:135–141.

———. 1970. Observaciones botánicas y geobotánicas en la costa del Caribe. *Revta Acad. colomb. Cienc. exact. fís. nat.* 13:415–465.

Hunziker, J. H.; Palacios, R. A.; de Valesi, A. G.; and Poggio, L. 1973. Species disjunctions in *Larrea*: Evidence from morphology, cytogenetics, phenolic compounds and seed albumins. *Ann. Mo. bot. Gdn.* 59:224–233.

Johnson, A. W. 1968. The evolution of desert vegetation in western North America. In *Desert Biology*, ed. G. W. Brown, vol.1, pp. 101–140. New York: Academic Press.

Koeppen, W. 1923. *Grundriss der Klimakunde*. Berlin and Leipzig: Walter de Gruyter & Co.

Lahey, J. F. 1958. *On the origin of the dry climate in northern South America and the southern Caribbean*. Ph.D. dissertation, University of Wisconsin.

Lasser, T. 1971. Zygophyllaceae. In *Flora de Venezuela, III*, ed. T. Lasser. Caracas: Instituto Botánico.

Loveless, A. R., and Asprey, C. F. 1957. The dry evergreen formations of Jamaica I. The limestone hills of the south coast. *J. Ecol.* 45:799–822.

Lowe, C.; Morello, J.; Goldstein, G.; Cross, J.; and Neuman, R. 1973. Análisis comparativo de la vegetación de los desiertos subtropicales de Norte y Sud América (Monte-Sonora). *Ecologia* 1:35–43.

McGinnies, W. G.; Goldman, B. J.; and Paylore, P. 1968. *Deserts of the world, an appraisal of research into their physical and biological environments*. Tucson: Univ. of Ariz. Press.

Marcuzzi, G. 1956. Contribución al estudio de la ecologia del medio xerófilo Venezolano. *Boln Fac. Cienc. for.* 3:8–42.

Meigs, P. 1953. World distribution of arid and semiarid homoclimates. In *Reviews of research on arid zone hydrology*, 1:203–209. Paris: Arid Zone Programme, Unesco.

Morello, J. 1958. La provincia fitogeográfica del Monte. *Op. lilloana* 2:1–155.

Pittier, H. 1926. *Manual de las plantas usuales de Venezuela*. 2d ed. Caracas: Fundación Eugenio Mendoza.

Pittier, H.; Lasser, T.; Schnee, L.; Luces de Febres, Z.; and Badillo, V. 1947. *Catálogo de la flora Venezolana*. Caracas: Litografía Vargas.

Porras, O.; Andressen, R.; and Pérez, L. E. 1966. *Estudio climatológico de las Islas de Margarita, Coche y Cubagua, Edo. Nueva Esparta*. Caracas: Ministerio de Agricultura y Cria.

Sarmiento, G. 1972. Ecological and floristic convergences between seasonal plant formations of tropical and subtropical South America. *J. Ecol.* 60:367–410.

————. 1973. The historical plant geography of South American dry vegetation. I. The distribution of the Cactaceae. Unpublished.

Shreve, F. 1951. Vegetation of the Sonoran desert. *Publs Carnegie Instn* 591:1–178.

Smith, L. B. 1971. Bromeliaceae. In *Flora de Venezuela, XII*, ed. T. Lasser. Caracas: Instituto Botánico.

Smith, R. F. 1972. La vegetación actual de la región Centro Occidental: Falcón, Lara, Portuguesa y Yaracuy de Venezuela. *Boln Inst. for lat.-am. Invest. Capacit.* 39–40:3–44.

Stebbins, G. L., and Major, J. 1965. Endemism and speciation in the California flora. *Ecol. Monogr.* 35:1–35.

Stehlé, H. 1945. Los tipos forestales de las islas del Caribe. *Caribb. Forester* 6:273–416.

Stoffers, A. L. 1956. The vegetation of the Netherlands Antilles. *Uitg. natuurw. Stud-Kring Suriname* 15:1–142.

Tamayo, F. 1941. Exploraciones botánicas en la Peninsula de Paraguaná, Estado Falcón. *Boln Soc. venez. Cienc. nat.* 47:1–90.

————. 1967. El espinar costanero. *Boln Soc. venez. Cienc. nat.* 111: 163–168.

Thornthwaite, C. W. 1948. An approach toward a rational classification of climate. *Geogr. Rev.* 38:155–194.

Trujillo, B. 1966. *Estudios botánicos en la región semiárida de la Cuenca del Turbio, Cejedes Superior*. Mimeographed.

Willis, J. C. 1966. *A dictionary of the flowering plants and ferns*. Cambridge: At the Univ. Press.

5. Adaptation of Australian Vertebrates to Desert Conditions   A. R. Main

Introduction

It is an axiom of modern biology that organisms survive in the places where they are found because they are adapted to the environmental conditions there. Current thinking has often associated the more subtle adaptations with physiological attributes, and the analysis of physiology has been widely applied to desert-dwelling animals in order to better understand their adaptation. Results of these inquiries frequently do not produce complete or satisfying explanations of why or how organisms survive where they do, and it is possible that explanations couched in terms of physiology alone are too simplistic. Clearly, while physiology cannot be ignored, other factors, including behavioral traits, need to be taken into account.

Accordingly, this paper sets out to interpret the adaptations of Australian vertebrates to desert conditions in the light of the physiological traits, the species ecology, and the geological and evolutionary history of the biota. To the extent that the components of the biota are integrated, its evolution can be conceived of as analogous to the evolution of a population; thus migrations and extinctions are analogous to genetic additions and deletions; and change in the ecological role of a component of the biota, the analogue of mutation.

The biota has changed and evolved mainly as a result of (a) changes in location and disposition of the land mass, (b) changes in the environment consequent on (a) above, and (c) extinctions and accessions. In the course of these changes strategies for survival will also change and evolve. It is the totality of these strategies which constitutes the adaptations of the biota.

Change in Location of Australia

The present continent of Australia appears to have broken away from East Antarctica in late Mesozoic times and to have moved to its present position adjacent to Asia in middle Tertiary (Miocene) times. In the course of these movements southern Australia changed its latitude from about 70°S in the Cretaceous to about 30°S at present (Brown, Campbell, and Crook, 1968; Heirtzler et al., 1968; Le Pichon, 1968; Vine, 1966, 1970; Veevers, 1967, 1971).

Changes in Environment

Prior to the fragmentation discussed above, the tectonic plate that is now Australia probably had a continental-type climate except when influenced by maritime air. As movement to the north proceeded, extensive areas were covered by epicontinental seas, and, later, extensive fresh-water lake systems developed in the central parts of the present continent. As Australia changed its latitude, the continental climate was influenced by the temperature of the surrounding oceans and particularly the temperature, strength, and origin of the ocean currents which bathed the shores. The ocean currents would in turn be driven by the global circulation, and the variations in the strength of the circulation and its cellular structure have affected not only the strength of the currents but also the climate of the continent. Frakes and Kemp (1972) suggested that for these reasons the Oligocene was colder and drier than the Eocene.

The present location of Australia across the global high-pressure belt, coupled with the fact that ocean currents driven by the west wind storm systems pass south of the continent, has meant the inevitable drying of the central lake systems and the onset of desert conditions in the interior of the continent. In the absence of marine fossils or volcanicity the precise timing of the stages in the drying of the continent is not possible. Stirton, Tedford, and Miller (1961, p. 23; and see also Stirton, Woodburne, and Plane, 1967) used the morphological evolution shown by marsupial fossils to infer possible age in terms of Lyellian epochs of the sedimentary beds in which marsupial fossils have been found. Ride (1971) tabulated the fossil evidence as it relates to macropods.

Two other events associated with the changed position of the continent have occurred concurrently: (a) the development of weathering profiles, especially duricrust formation, on the land surface; and (b) changes in the composition of the flora.

Weathering profiles capped with duricrust are widespread throughout Australia, and Woolnough (1928) believed this duricrust to be synchronously developed over an enormous area. Since the Upper Cretaceous Winton formation was capped by duricrust, Woolnough believed the episode to be of Miocene age. The climate at this time of peneplanation and duricrust formation was thought to be marked by well-defined wet and dry seasons, so that the more soluble material was leached away in the wet season, and less soluble and particularly colloidal fraction of the weathering products was carried to the surface and precipitated during the dry season. Recent work in Queensland where basalts overlie deep weathering profiles indicates that deep weathering took place earlier than early Miocene (Exon, Langford-Smith, and McDougall, 1970). Other workers suggested that, as the climate becomes progressively drier, weathering processes follow a sequence from laterite formation through silcrete formation to aeolian processes and dune formation (Watkins, 1967).

Biologically the significant aspect of duricrust formation is, however, the removal from, or binding within, the weathering profile of soluble plant nutrients. Beadle (1962a, 1966) showed experimentally that the woodiness which is so characteristic of Australian plants is to some extent related to the low phosphorus status of the soil. Australian soils are well known for their low phosphorus status (Charley and Cowling, 1968; Wild, 1958). It has been argued that the low phosphorus is due to the low status of the parent rocks (Beadle, 1962b) or to the leaching which occurred during the process of laterization (Wild, 1958).

Changes in the floral composition are indicated by the fossil and pollen record. Early in the Tertiary, pollen of southern beech (*Nothofagus*), in common with other pollen present in these deposits, suggests that a vegetation with a floral composition similar to that of present-day western Tasmania was widespread in southern Australia (see fig. 5-1), for example, at Kojonup (McWhae et al., 1958); Cool-gardie (Balme and Churchill, 1959); Nornalup, Denmark, Pidinga, and Cootabarlow, east of Lake Eyre (Cookson, 1953; Cookson and Pike, 1953, 1954); and near Griffith in New South Wales (Packham, 1969, p. 504).

Later the Lake Eyre deposits show a change, and the pollen record is dominated by myrtaceous and grass pollen (Balme, 1962). By Plio-cene times the fossil record is restricted to eastern Australia and sug-gests a cool rain forest with *Dacrydium, Araucaria, Nothofagus*, and *Podocarpus*, which was later replaced by wet sclerophyll forest with *Eucalyptus resinifera* (Packham, 1969, p. 547). This record is consis-tent with a drying of the climate; however, in Tasmania comparable changes in the floral composition—that is, from *Nothofagus* forest to myrtaceous shrub or *Eucalyptus* woodland with a grass understory—result from fire (Gilbert, 1958; Jackson, 1965, 1968a, 1968b), and it seems highly likely that associated with the undeniable deterioration of the climate there occurred an increased incidence of fire.

Many authors have recognized that the present Australian flora not only is adapted to periodic fires but also includes many species which are dependent on fire for their persistence (Gardner, 1944, 1957; Mount, 1964, 1969; Cochrane, 1968). At present many wild or bush fires are intentionally lit or are the result of man's carelessness, but every year there are many fires which are caused by lightning strike (Wallace, 1966).

Fires are important in the Australian arid, semiarid, and seasonally arid environments because it is principally from the ash beds resulting from intense fires, and not from the slow decay of plant material, that nutrients are returned to the soil. It is thought that the way natural of the common Australian shrubs and trees and their fire dependence reflect an evolutionary adaptation to fire. There is no doubt that in the past, in the absence of man, many intense fires were lit when lightning strike ignited ample and highly inflammable fuel.

Apart from returning nutrients to the soil, fire appears to be an important ecological factor in habitats ranging from the well-watered coastal woodlands dominated by *Eucalyptus* forests to the hummock grassland (dominated by *Triodia*) of the arid interior (Burbidge, 1960; Winkworth, 1967). Numerous postfire successions occur depending on the season of the burn, the quantity of fuel, and the frequency of burning. As an ecological factor, in arid Australia fire is as ubiquitous as drought.

Not all the changes in the biota have been due to fire and the increasing aridity. Numerous elements of the flora must have invaded Australia and then colonized the arid sandy interior by way of littoral sand dunes (Gardner, 1944; Burbidge, 1960). This invasion of Australia could only have occurred after the collision of the Australian plate with Asia in Miocene times. Simultaneously these migrant plant species would have been accompanied by rodents and other vertebrates of Asian affinities which also invaded through similar channels (Simpson, 1961).

To summarize the foregoing, Australia arrived in its present position from much higher southern latitudes, and the change in latitude was associated with a change in climate which passed from being mild and uniform in early Tertiary through marked seasonality to severe and arid by the end of the Pleistocene. Associated with climatic change two things occurred: first, a removal of plant nutrients and the probable development of a "woody" flora, and, second, the concurrent appearance of fire as a significant ecological factor.

Extinctions and Accessions

The climatic changes led to numerous extinctions in the old vertebrate fauna, for example, the Diprotodontidae; to a marked development in macropod marsupials (Stirton, Tedford, and Woodburne, 1968; Woodburne, 1967), which are adapted to the low-nutrient-status fibrous plants; and to the radiation of those Asian invaders which could exploit the progressive development of an arid climate in central Australia. As a result of the events outlined above, the fauna of arid Australia consists of two elements:

1. An older one originating in a cool, high-latitude climate now adapted to or at least persisting under arid conditions. Ride (1964), in his review of fossil marsupials, placed *Wynyardia bassiana*, the oldest diprotodont marsupial known, as of Oligocene age. This was at a time when Australia still occupied a southern location far distant from Asia (Brown et al., 1968, p. 308), suggesting that marsupials are part of the old fauna not derived from Asia.

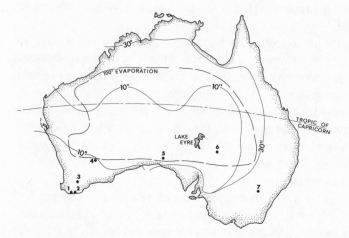

Fig. 5-1. Map of Australia showing approximate extent of arid zone as defined in Slatyer and Perry (1969). The northern boundary corresponds to the 30-inch (762 mm) isohyet, while the southern boundary corresponds with the 10-inch (242 mm) isohyet. The northern boundary of the 10-inch isohyet is also shown, as is the approximate boundary of the region experiencing evaporation of 100 inches (2,540 mm) or more per year. Localities from which fossil pollen recorded: 1. Nornalup; 2. Denmark; 3. Kojonup; 4. Coolgardie; 5. Pidinga; 6. Cootabarlow; 7. Lachlan River occurrence (near Griffith).

2. A younger element (not older than the time at which Australia collided with Asia) derived by evolution from migrants which established themselves on beaches. Rodents, agamid lizards, elapid snakes, and some bird groups fall in this category. These invasive episodes have continued up to the present.

The Australian Arid Environment

The extent of the Australian arid zone is shown in figure 5-1. This area is characterized by irregular rainfall, constant or seasonal shortage of water, high temperatures, and, for herbivores, recurrent seasonal inadequacy of diet. In many respects the arid area manifests in a more intense form the less intense seasonal or periodic droughts of the surrounding semiarid areas. Earlier it has been suggested that the present arid conditions are merely the terminal manifestation of climatic conditions which commenced to deteriorate in middle Tertiary times.

Desert Adaptation

A biota which has survived under increasingly arid conditions for such a long period of time might be expected to have evolved well-marked adaptations to high temperatures, shortage of water, and poor-quality diet. Yet one would not necessarily expect that all species would show similar or equal adaptive responses. The reason for this is that the incidence of intense drought is patchy; and, while some parts of arid Australia are always suffering drought, the areas suffering drought in the same season or between seasons are spatially discontinuous and often widely separated, so that it is conceivable that a mobile population could flee drought-stricken areas for more equable parts. Moreover, since fire as well as drought is ubiquitous in arid Australia, the benefits resulting from break of drought may be quite different according to whether an area has been recently burnt or not. Furthermore, these differences will be dissimilar depending on whether the biotic elements occur early or late in seral stages of the postfire succession. Indeed, many animal species which occupy the late seral or climax stages of the postfire succession could conceivably avoid much of the heat stress and resultant water shortage consequent upon evaporative cooling by behaviorally seeking out the cooler sites in the climax vegetation. All of the foregoing suggest that for the Australian arid biota we should not only look at the stressful factors (high temperature, water shortage, and quality of diet) but also determine the postfire seral stages occupied by the species.

Tolerance to the stressful factors is important because it affects the individual's ability to survive to reproduce. When individuals of a population reproduce successfully, the population persists. However, in its persistence a population will not maintain constant numerical abundance, because, for example, drought and fire will reduce numbers; and species favoring one seral stage of the postfire succession will, in any one locality, vary from being rare, then abundant, and finally again rare, as the preferred seral stage is passed through. For any species only detailed inquiry will show how population characteristics of reproductive capacity, age to maturity, longevity, and dispersal are related to the individual's ability to survive drought.

Individual Responses

Mammals

Among mammals, and marsupials especially, the macropods (kangaroos and wallabies) have received most study, but some work has been done on rodents, particularly *Notomys* (MacMillan and Lee, 1967, 1969, 1970). Both these groups are herbivores, and the macropods particularly show spectacular adaptations to the fibrous nature of their food plants and the attendant low-nutrition quality of the diet.

With the exception of the forest-dwelling *Hypsiprimnodon moschatus*, all kangaroos and wallabies so far investigated have a ruminantlike digestive system. This is an especially elaborated saccular development of the alimentary tract anterior to the true stomach. Within the sacculated "ruminal area" the fibrous ingesta are retained and fermented by a prolific bacterial flora and protozoan fauna. As a result of this activity, otherwise indigestible cellulose is broken down to material which can be metabolized by the kangaroo as an energy source (Moir, Somers, and Waring, 1956).

The bacteria of the gut need a nitrogen source in order to grow. In general, most natural diets in arid Australia are low in nitrogen; however, the bacteria of the kangaroo's gut are able to use urea as a nitrogen source (Brown, 1969) and so can supplement dietary nitrogen by recycling urea which would otherwise require water for its excretion in urine. The common red kangaroo *Macropus rufus* (the euro) can remain in positive nitrogen balance on a diet which contains less than 1 g nitrogen per day for a euro of the average body weight of 12.8 kg (Brown and Main, 1967; Brown, 1968). However, it can supplement the dietary intake of nitrogen by recycling urea (Brown, 1969). In this connection the two common kangaroos (table 5-1) of arid Australia show contrasting solutions to the stresses imposed by heat, drought, and shortage of water, and in fact respond differently to seasonal stress (Main, 1971).

Body temperatures of marsupials are elevated at high ambient temperatures, but they are maintained below environmental tempera-

Table 5-1. *Comparison of Adaptations of Two Arid-Land Species of Kangaroos*

Characteristic	Red Kangaroo (*Megaleia rufa*)	Euro (*Macropus robustus*)
Coat	short, close; highly reflective	long, shaggy; not reflective
Preferred shelter	sparse shrubs	caves and rock piles
Diet	better fodder; higher nitrogen content	poorer fodder; lower nitrogen content
Temperature regulation	reflective coat; evaporative cooling	cool of caves or rock piles; evaporative cooling
Water shortage	acute shortage not demonstrated	unimportant except when shelter inadequate
Urea recycling	not pronounced; urea always high in urine	pronounced when fermentable energy as starch available, e.g., as seed heads of grasses
Electrolyte	higher in diet	lower in diet
Population characteristics	flock; locally nomadic	solitary; sedentary
Breeding	continuous	continuous

Source: Data from Dawson and Brown (1970), Storr (1968), and Main (1971).

tures. Temperature regulation under hot conditions appears to be costly in terms of water (Bartholomew, 1956; Dawson, Denny, and Hulbert, 1969; Dawson and Bennet, 1971).

MacMillan and Lee (1969) studied two species of desert-dwelling *Notomys* and interpreted their findings in terms of the contrasting habitats occupied, so that the salt-flat-dwelling *Notomys cervinus* has a kidney adapted to concentrating electrolytes while the sandhill-dwelling *N. alexis* was adapted to concentrating urea.

Birds

Because of their mobility, birds in an arid environment present a different set of problems to those presented by both mammals and lizards. They can and do fly long distances to watering places and, when water is available, may use it for evaporative cooling. Many of the adaptations are likely to be related to conservation of water so that the frequency of drinking is reduced. Fisher, Lindgren, and Dawson (1972) studied the drinking patterns of many species, including the zebra finch and budgerigar which have been shown under laboratory conditions to consume little or no water (Cade and Dybas, 1962; Cade, Tobin, and Gold, 1965; Calder, 1964; Greenwald, Stone, and Cade, 1967; Oksche et al., 1963).

It is likely that the ability of the budgerigar and the zebra finch to withstand water deprivation in the laboratory reflects their ability to survive in the field with minimum water intake and so exploit food resources which are distant from the available water. Johnson and Mugaas (1970) showed that both these species possess kidneys which are modified in a way which assists in water conservation.

Reptiles

Four responses of individuals to hot arid environments are readily measured: preferred temperatures, heat tolerance, rates of pulmonary and cutaneous water loss, and tolerance to dehydration.

Under field conditions nocturnal lizards, for example, geckos, have to tolerate the temperatures experienced in their daytime shelters. On the other hand, diurnal species, for example, agamids, such as *Amphibolurus*, have body temperatures higher than ambient temperatures during the cooler parts of the year and body temperatures cooler than ambient during the hotter season. The body temperatures recorded for field-caught animals indicate a specific constancy (Licht et al., 1966b) which comes about by a series of behavioral responses which range from body posture to avoidance reactions (Bradshaw and Main, 1968). A large series of data on body temperatures in the field has been presented by Pianka (1970, 1971a, and 1971b) and Pianka and Pianka (1970).

With the exception of *Diporophora bilineata*, with a mean body temperature in the field of 44.3°C (Bradshaw and Main, 1968), no species recorded a mean temperature above 39°C. However, arid-land species spend more of their time in avoidance reactions than do species from semiarid situations (Bradshaw and Main, 1968). Further information on preferred body temperature can be obtained by placing lizards in a temperature gradient and allowing them to choose a body temperature.

Data from neither the field-caught animals nor those selecting temperature in a gradient indicate any marked preference for exceptionally high temperatures on the part of most lizard species. However, in a situation where choice of temperature was not possible, it is conceivable that species from arid environments could tolerate higher temperatures for a longer period than species from less-arid situations. Bradshaw and Main (1968) compared *Amphibolurus ornatus*, a species from semiarid situations, with *A. inermis*, a species from arid areas, after acclimating them to 40°C and then exposing them to 46°C. Their mean survival times were 64 ± 5.6 and 62 ± 6.58 minutes respectively. There was no statistically significant difference in the survival time of each species. These results suggest that the major adaptation of *Amphibolurus* species to hot arid environments is likely to be the development of a pattern of behavioral avoidance of heat stress.

Not all lizards in hot arid situations show the pattern of *Amphibolurus* sp. and *Diporophora bilineata*, which when acclimated to 40°C can withstand an exposure of six hours to a body temperature of 46°C without apparent ill effect and survive for thirty minutes at 49°C (Bradshaw and Main, 1968). The nocturnal geckos, which must tolerate the temperature of their daytime refuge, show another pattern illustrated by *Heteronota binoei*, a species sheltering beneath litter; *Rhynchoedura ornata*, a species which frequently shelters in cracks and holes (deserted spider burrows) in bare open ground; and *Gehyra variegata*, a species sheltering beneath bark. Data for these three species are given in table 5-2. These data suggest that adaptation to high temperatures in Australian geckos is considerably modified by behavioral and habitat preferences and is not directly related to increased aridity in a geographical sense.

Bradshaw (1970) determined the respiratory and cutaneous components of the evaporative water loss in specimens of *Amphibolurus ornatus*, *A. inermis*, and *A. caudicinctus* (matched for body weight) held in the dry air at 35°C after being held under conditions which allowed them to attain their preferred body temperature by behavioral regulation. Bradshaw showed that evaporative water loss was greatest in *A. ornatus* and least in *A. inermis*. He also showed that losses by both pathways were reduced in the desert species. All differences were statistically significant. However, while the cutaneous component was greater than the respiratory in *A. ornatus*, it was less than the respiratory in *A. inermis*. Bradshaw also compared CO_2 production of uniformly acclimated *A. ornatus* with that of *A. inermis* and showed that CO_2 production and respiratory rate of *A. inermis* were significantly lower than *A. ornatus*. Bradshaw concluded that the greater water economy of desert-living *Amphibolurus* was achieved both by reduction in metabolic rate and change to a more impervious integument.

By means of a detailed field population study of *A. ornatus*, Bradshaw (1971) was able to show that individuals of the same cohort grew at different rates so that some animals matured in one, two, or three years. These have been referred to as fast- or slow-growing animals. Bradshaw, using marked animals of known growth history, showed that during summer drought there was a difference between fast- and slow-growing animals with respect to distribution of fluids and electrolytes. Slow-growing animals showed no difference when compared with fully hydrated individuals except that electrolytes in plasma and skeletal muscle were elevated. Fast-growing animals, however, showed weight losses and changes in fluid volume. Weight losses greater than 20 percent of hydrated weight encroached upon the extracellular fluid volume; but the decrease in volume was restricted to the interstitial fluid, leaving the circulating fluid volume intact, that is, the blood volume and plasma volume remained constant. Earlier, Bradshaw and Shoemaker (1967) showed that the diet of *A. ornatus* consisted of sodium-rich ants and that during summer the lizards lacked sufficient water to excrete the electrolytes without using body water. Instead, the sodium ions were retained at an elevated level in the extracellular fluid which increased in volume by an isosmotic shift of fluid from the intracellular compartment. This sodium retention operates to protect fluid volumes when water is scarce and so enhances survival. Electrolytes were excreted following the occasional summer thunderstorm.

In his population study Bradshaw (1971) showed that only fast-growing animals died as a result of summer drought. Bradshaw (1970) extended his study of water and weight loss in field populations to other species including *A. inermis* and *A. caudicinctus*. As a result of this study he concluded that only males of *A. inermis* lost weight and that, in all species studied, fluid volumes were protected by the retention of sodium ions during periods when water was short. Bradshaw also showed that sodium retention occurred in *A. ornatus* in midsummer but only occurred in *A. inermis* and *A. caudicinctus* after long and intense drought.

Both *A. inermis* and *A. caudicinctus* complete their life cycle in a year (Bradshaw, 1973, personal communication; Storr, 1967). They are thus fast growing in the classification used to describe the life history of *A. ornatus*; but, either by a change in metabolic rate and integument or by some other means, they have avoided the deleterious effects associated with the rapid development of *A. ornatus*.

Population Response

The capacity and speed with which a species can occupy an empty but suitable habitat are related to its capacity to increase. Cole (1954) pointed out that time to maturity, litter size, and whether reproduction is a single episode or repeated throughout the female's life bear on the rate at which a population can increase; but he believed that reproduction early in life was most important for the population. No systematic recording of life-history data appears to have been undertaken in Australia, but such information is critical for understanding how populations persist in fluctuating environments. Whether the fluctuations are due to recurrent drought or to fire or seral stages of postfire succession is not too important, because following any of these events a population nucleus will have the opportunity to expand quickly into an empty but suitable habitat. Moreover, its chances of persisting are enhanced if it can very quickly occupy all favorable habitats at the maximum density because the random spatial distribution of the next drought or fire sequence will determine the sites of the next *refugium*.

The foregoing would suggest that modification of the life history, particularly early maturity, might be as important as physiological adaptation under Australian arid conditions. However, young or small animals are at a disadvantage because of the effects of metabolic body size compared with larger, older mature animals, and hence there is an advantage in late maturity and greater longevity, so that the risks of death which are related to metabolic body size are spread more favorably than they would be in a species in which each generation lived for only one year. Undoubtedly, natural selection will have produced adaptations of life history so that the foregoing apparent conflicts are resolved.

Several workers—MacArthur and Wilson (1967) and Pianka (1970)—have considered the response of populations to selection in terms of whether high fecundity and rapid development or individual fitness and competitive superiority have been favored. These two types of selection were referred to by MacArthur and Wilson (1967) as r-selection and K-selection. Pianka (1970) has tabulated the correlates of each type of selection.

King and Anderson (1971) pointed out that, if a cyclically changing environment varies over few generations, r-type selection factors will be dominant; on the other hand, in a changing environment which has a period of fluctuation many generations long, K-type selection will be dominant. In this connection we might consider early maturity and large clutch size as manifesting the response to r-type selection; and slower maturity, smaller clutch size, well-marked display, and other devices for marking territory as representing responses to K-type selection.

Mertz (1971) showed that the response of a population to selection will be different depending on whether the population is increasing or declining. In the latter case selection favors the long-lived individual which continues to breed and is thus able to exploit any environmental amelioration even if it occurs late in life. This type of selection tends to produce long-lived populations.

Earlier it was suggested that Australia has been subjected to a prolonged climate and fire-induced deterioration of the environment which might be expected to produce a response akin to that envisaged by Mertz (1971) and unlike the advantageous rapid development and early reproduction mentioned by Cole (1954). Selection for longevity is a special case in which competitive superiority is principally expressed in terms of a long reproductive life. Murphy (1968) showed this was as a consequence of uncertainty in survival of the prereproductive stages.

With the foregoing outline, it is possible to consider the little information known about the life histories of species from arid Australia in terms of whether they reflect selection during the past for capacity to increase, competitive efficiency related to carrying capacity, or longevity.

Mammals

Macropods. The fossil record suggests that in both Tertiary and post-Pleistocene times the macropods have increased their dominance of the fauna despite the general deterioration of the climate (Stirton et al., 1968). It has already been suggested that the ruminantlike digestion preadapted these species to the desert conditions. The highly developed ruminantlike digestion of macropods can be viewed as a device for delaying the death from starvation caused by a nutritionally inadequate diet. It is thus a device for maximizing physiological longevity once adulthood is achieved.

Among the marsupials there is considerable diversity in their life histories, but there appear to be tendencies toward longevity with respect to populations in arid situations as indicated in the two cases below:

1. In the typical mainland swampy situations the quokka (*Setonix brachyurus*) matures early and breeds continuously and is apparently not long lived. On the other hand, a population of this species on the relatively arid Rottnest Island is older than the mainland form when it first breeds. Breeding is seasonal, and so Rottnest animals tend to produce fewer offspring per unit time than the mainland form (Shield, 1965). Moreover, individuals from the island population tend to live seven to eight years, with a few females present and

still reproducing in their tenth year. The pollen record on Rottnest indicates that the environment has declined from a woodland to a coastal heath and scrubland over the past 7,000 years (Storr, Green, and Churchill, 1959). Despite the difference in detail, the modification in the life history of the quokka on the semiarid Rottnest Island achieves the same end as the red kangaroo and euro discussed below.

2. Typical arid-land species, such as the red kangaroo, *Megaleia rufa*, and the euro, *Macropus robustus*, have no defined season of breeding, and females are always carrying young except under very severe drought. Both these species tend to be long lived (Kirkpatrick, 1965), and females may still be able to bear young when approaching twenty years of age.

The breeding of the red kangaroo and euro suggests that adaptation of life history has centered around the metabolic advantages of large body size in a long-lived animal which is virtually capable of continuous production of offspring, some of which must by chance be weaned into a seasonal environment which permits growth to maturity.

Numerous workers have shown that macropod marsupials have lowered metabolic rates with which are associated reduced requirements for water, energy, and protein and a slower rate of growth. The first three of these are of advantage during times of drought; and, should the last contribute to longevity, it will also be advantageous, insofar as offspring have the potential to be distributed into favorable environments whenever they occur.

Rodents. The Australian rodents appear to have a typical rodent-type reproductive pattern with a high capacity to increase. They appear to be able to survive through drought because of their small size and capacity to persist as small populations in minor, favorable habitats.

Birds

Most bird species which have been studied physiologically belong to taxonomic groups which also occur outside Australia, for example, finches, pigeons, caprimulgids, and parrots. The information on which a comparative study of modifications of the life histories of the Australian forms with their old-world relatives could be based has not been assembled. However, several observations—for example, Cade et al. (1965), that Australian and African estrildine finches are markedly different in physiology, and Dawson and Fisher (1969), that the spotted nightjar (*Eurostopodus guttatus*), like all caprimulgids, has a depressed metabolism—are suggestive that the life histories of some Australian species (finches) might be highly modified, while others show only slight modification from their old-world relatives (caprimulgids); and these may, in a sense, be thought of as being preadapted to survival in arid Australia.

Keast (1959), in a review of the life-history adaptations of Australian birds to aridity, showed that the principal adaptation is opportunistic breeding after the break of drought when the environment can provide the necessities for successful rearing of young. Longevity of individuals in unknown, but the breeding pattern is consistent with selection which has favored longevity.

Fisher et al. (1972) observed that honeyeaters (Meliphagidae), which are widespread and common throughout arid Australia, are surprisingly dependent on water. The growth of these birds to maturity and their metabolism are not known, but these authors speculated that the dependence may be due in large part to the water loss attendant on the activity associated with the high degree of aggressive behavior exhibited by all species of honeyeaters.

The following speculation would be consistent with the observations of Fisher et al. (1972): Most honeyeaters frequent late and climax stages when the vegetation is at its maximum diversity with numerous sources of nectar and insects. Such a habitat preference would suggest that K-type selection would have operated in the past, and the advantages of obtaining and maintaining an adequate territory by aggressive display may outweigh any disadvantages of individual high water needs which were consequent upon the aggressive display.

Reptiles

Table 5-2 has been compiled from the information available on lizard physiology and biology. The information is not equally complete for all species tested; however, it does suggest that Australian desert

Table 5-2. *Physiological, Ecological, and Life-History Information for Selected Species of Australian Lizards*

Species	Mean Preferred Temperature	Mean Survival		Water Loss (mg/g/hr)	Seral Stage
		Minutes	Temperature (°C)		
Amphibolurus inermis	36.4[a]	102.8 2.0	46.0 48.0	1.05 at 35°C	burrows in early seres
A. caudicinctus	37.7[a]	92.8 45.0	46.0 47.0	1.80 at 35°C	rock piles in climax hummock grassland
A. scutulatus	38.2[a]	40.8 28.0	46.0 47.0	?	shady climax woodland
Diporophora bilineata	44.3[b]	360.0 29.5	46.0 49.0	?	fire disclimax
Moloch horridus	36.7[a]	?	?	?	late seres and climax

Age to Maturity (yrs)	Reproduction		Longevity (yrs)	Reference
	No. of Clutches per Year	Eggs per Clutch (means)		
0.75	possibly 2	3.43	1	Licht et al., 1966*a*, 1966*b*; Pianka, 1971*a*; Bradshaw, personal communication
0.75	possibly 2	?	1	Licht et al., 1966*a*, 1966*b*; Storr, 1967; Bradshaw, 1970
?	possibly only 1	6.5	?	Licht et al., 1966*a*, 1966*b*; Pianka, 1971*c*
?	?	?	?	Bradshaw and Main, 1968
3–4	usually 1	6–7	6–20	Sporn, 1955, 1958, 1965; Licht et al., 1966*b*; Pianka and Pianka, 1970

Species	Mean Preferred Temperature	Mean Survival		Water Loss (mg/g/hr)	Seral Stag
		Minutes	Temperature (°C)		
Gehyra variegata	35.3[a]	72.8 2.0	43.5 46.0	2.07 at 25°C 3.37 at 30°C 3.80 at 35°C	climax anc postclima> woodland
Heteronota binoei	30.0[ac]	162.0 0.0	40.5 43.5	0.27 at 30°C	climax wit litter
Rhynchoe-dura ornata	34.0[a]	55.3	46.0	?	holes in b soil in clin woodland

[a]In gradient.   [b]In field.   [c]May be too high—see Licht et al., 1966*b*.

species exhibit a wide range of tolerances to elevated temperatures. It is surprising, for example, that *Heteronota binoei* survives at all in the desert. Geckos, depending on the species, may have clutches of a single egg, but no species have clutches larger than two eggs; however, they may have one or two clutches each breeding season, *Heteronota binoei* and *Gehyra variegata* have respectively one and two clutches. *Heteronota binoei*, with an apparent preference for low temperatures and an inability to tolerate high temperatures, has adapted to the desert by its extremely low rate of water loss, behavioral attachment to sheltered climate situations, and, relative to *G. variegata*, early maturity and large clutch size (two eggs vs. one).

On the other hand, *G. variegata* is better adapted to high temperatures and, even though it is relatively poor at conserving water, is able to survive in the deteriorating and more exposed situations of the late climax and postclimax. Moreover, these physiological adapta-

| Age to Maturity (yrs) | Reproduction | | Longevity (yrs) | Reference |
	No. of Clutches per Year	Eggs per Clutch (means)		
2; breed in 3rd	2	1	mean 4.4	Bustard and Hughes, 1966; Licht et al., 1966*a*, 1966*b*; Bustard, 1968*a*, 1969; Bradshaw, personal communication
1.6 or 2.5	usually 1	2	mean 1.9	Bustard and Hughes, 1966; Licht et al., 1966*a*; Bustard, 1968*b*
?	?	?	?	Licht et al., 1966*a*, 1966*b*

tions are associated with a long adult life and thus enhance the possibility of favorable recruitment in any season where conditions are ameliorated so that eggs and young have an enhanced survival.

Among the agamids the information is not nearly as complete. *Amphibolurus inermis* and *A. caudicinctus* are early maturing, short-lived species relying on a high rate of reproduction to maintain the population and are thus the analogue of *H. binoei*. *Moloch horridus* and *A. scutulatus*, on the other hand, appear to be the analogue of *G. variegata*; and it is unfortunate that information on age to maturity and longevity of *A. scutulatus* is not available. One can only speculate on age to maturity and longevity of *Diporophora*, but it seems likely that recruitment would only be successful in years when summer cyclonic rain ameliorated environmental conditions; and one might guess that it is a long-lived animal.

It is interesting that the fast-maturing species either have a cool refuge in which the small young can establish themselves (*H. binoei* in climax) or a cool season in which they can grow to almost adult size (*A. inermis*, *A. caudicinctus*). In addition, these species have another adaptation in producing twin broods in each breeding season. Should there be a drought, the young from the first brood will almost certainly be lost. However, should the young be born into a season in which thunderstorms are common, they would be able to thrive under almost ideal conditions. Since the offspring from the second clutch are born late in the summer or early autumn, they are almost certain to survive regardless of the preceding summer conditions.

Discussion

The foregoing suggests that early in Tertiary times Australia underwent a change in position from higher (southern) to lower (tropical) latitudes. Stemming from this there has been a prolonged and disastrous change in climate toward increasing aridity. This has been accompanied by the increased incidence of fire as an ecological factor.

Much of the original biota has become extinct as the result of these changes, but there have been some additions from Asia. Both the old and new elements of the biota that have survived to the present have done so because they have been able to accommodate their individual physiology and population biology to the stresses imposed by climatic deterioration (drought) and fire.

The foregoing has been achieved by a series of complementary strategies as follows:

1. Physiological strategies
    a. Behavioral avoidance of stressful environmental factors
    b. Heat tolerance
    c. Ability to conserve water, including ability to handle electrolytes
    d. Ability to survive on diets of low nutritional value (herbivores)
2. Reproductive strategies
    a. High reproductive capacity, so enabling a population nucleus surviving after drought to rapidly repopulate the former range and to occupy all areas which could possibly form *refugia* in future droughts

b. Increased competitive advantage by means of small well-tended broods of young and well-developed displays for holding territories

c. Increased longevity, so that adults gain advantage from metabolic body size while young are produced over a span of years, so ensuring that at least some are born into a seasonal environment in which they can survive and become recruits to the adult population

It is thus apparent that vertebrates inhabiting arid parts of Australia display a diversity of individual adaptations to single components of the arid environment, and it is difficult to interpret the significance of experimental laboratory findings achieved as the result of simple single-factor experiments. For example, under experimental conditions, kangaroos and wallabies, if exposed to high ambient temperatures, use quantities of water in evaporative cooling (Bartholomew, 1956; Dawson et al., 1969; Dawson and Bennet, 1971).

Yet these arid-land species are capable of surviving intense drought conditions when the environment provides the appropriate shelter conditions. These may be postfire seral stages as needed by the hare wallaby, *Lagorchestes conspicillatus* (Burbidge and Main, 1971), or rock piles needed by the euro, *Macropus robustus*. Given that the euro and hare wallaby have shelter of the appropriate quality, both species are apparently well adapted to grow and reproduce on the low-quality forage which is available where they live. Moreover, both species are capable of reproduction at all seasons so that, while their reproductive potential is limited—because of having only one young at a time—they do maximize their reproductive potential by continuous breeding and by distributing the freshly weaned young at all seasons, which is particularly important in a seasonally unpredictable environment.

In general, while it is true that some species show a highly developed degree of adaptation to arid conditions, it is difficult to find a case which is unrelated to seral successional stages. A pronounced example of this is afforded by the lizard *Diporophora bilineata* and the gecko *Diplodactylus michaelseni*, which can withstand higher field body temperatures than any other Australian species but which appear to be abundant only in excessively exposed fire disclimax situations.

Most of the vertebrates which survive in the desert appear to do so not solely because of well-developed individual adaptation (tolerance) to the hot dry conditions of arid Australia, but because of habitat preference and population attributes which permit the species to cope, first, with the ecological consequences of fire and, second, with drought. In a sense, adaptation to fire has preadapted the vertebrates to drought.

Desert species have had to choose whether the ability of a population to grow is equivalent to ability to persist. Two circumstances can be envisaged in which ability to grow is equivalent to persistence: when rapid repopulation of an area after drought will ensure that all potential future *refugia* are occupied and when rapid population growth excludes other species from a resource.

In the desert where drought conditions are the norm, however, persistence is achieved by females replacing themselves with other females in their lifetime. This requires that juveniles must withstand or avoid desert conditions until they reach reproductive age. Seasonal amelioration of conditions in desert environments is notoriously unpredictable, and it seems that many Australian desert animals persist as populations because of long reproductive lives during which some young will be produced and grow to maturity.

In considering the individual and the population aspects of survival we should envisage the space occupied by an animal as providing scope for minimizing the environmental stresses of heat, water shortage, and poor-quality diet. An animal will choose to live in places where the stresses are least; when these are not available, it will select sites or opt for physiological responses which allow it to prolong the time to death. Urea recycling by macropods should be viewed in this light. When environmental amelioration occurs, it is taken as an opportunity to repopulate the population by recruiting young.

Acknowledgments

Financial assistance is acknowledged from the University of Western Australia Research Grants Committee, the Australian Research Grants Committee, and Commonwealth Scientific and Industrial Research Organization. Professor H. Waring, Dr. S. D. Bradshaw, and Dr. J. C. Taylor kindly read and criticized the manuscript.

Summary

It is suggested that the Australian deserts developed as a consequence of the movement in Tertiary times of the continental plate from higher latitudes to its present position. An increasing incidence of wild fire is associated with the development of dry conditions. The vertebrate fauna has adapted to the development of deserts and incidence of fire at two levels: (a) the individual or physiological, emphasizing such strategies as behavioral avoidance of stressful conditions, conservation of water, tolerance of high temperatures, and, with macropod herbivores, ability to survive on low-quality forage and through the supplementation of nitrogen by the recycling of urea; and (b) the population, emphasizing reproductive strategies and longevity, so that young are produced over a long period of time thus enhancing the possibility of successful recruitment.

It is further suggested that survival of individuals and persistence of the population are only possible when the environment, especially the postfire plant succession, provides the space and scope for the implementation of the strategies which have evolved.

References

Balme, B. E. 1962. Palynological report no. 98: Lake Eyre no. 20 Bore, South Australia. In *Investigation of Lake Eyre*, ed. R. K. Johns and N. H. Ludbrook. *Rep. Invest. Dep. Mines S. Aust.* No. 24, pts. 1 and 2, pp. 89–102.

Balme, B. E., and Churchill, D. M. 1959. Tertiary sediments at Coolgardie, Western Australia. *J. Proc. R. Soc. West. Aust.* 42:37–43.

Bartholomew, G. A. 1956. Temperature regulation in the macropod marsupial *Setonix brachyurus*. *Physiol. Zoöl.* 29:26–40.

Beadle, N. C. W. 1962a. Soil phosphate and the delimitation of plant communities in Eastern Australia, II. *Ecology* 43:281–288.

———. 1962b. An alternative hypothesis to account for the generally low phosphate content of Australian soils. *Aust. J. agric. Res.* 13: 434–442.

———. 1966. Soil phosphate and its role in molding segments of the Australian flora and vegetation, with special reference to xeromorphy and sclerophylly. *Ecology* 47:992–1007.

Bradshaw, S. D. 1970. Seasonal changes in the water and electro-lyte metabolism of *Amphibolurus* lizards in the field. *Comp. Biochem. Physiol.* 36:689–718.

———. 1971. Growth and mortality in a field population of *Amphibolurus* lizards exposed to seasonal cold and aridity. *J. Zool., Lond.* 165:1–25.

Bradshaw, S. D., and Main, A. R. 1968. Behavioral attitudes and regulation of temperature in *Amphibolurus* lizards. *J. Zool., Lond.* 154:193–221.

Bradshaw, S. D., and Shoemaker, V. H. 1967. Aspects of water and electrolyte changes in a field population of *Amphibolurus* lizards. *Comp. Biochem. Physiol.* 20:855–865.

Brown, D. A.; Campbell, K. S. W.; and Crook, K. A. W. 1968. *The geological evolution of Australia and New Zealand*. Oxford: Pergamon Press.

Brown, G. D. 1968. The nitrogen and energy requirements of the euro (*Macropus robustus*) and other species of macropod marsupials. *Proc. ecol. Soc. Aust.* 3:106–112.

———. 1969. Studies on marsupial nutrition. VI. The utilization of dietary urea by the euro or hill kangaroo, *Macropus robustus* (Gould). *Aust. J. Zool.* 17:187–194.

Brown, G. D., and Main, A. R. 1967. Studies on marsupial nutrition. V. The nitrogen requirements of the euro, *Macropus robustus*. *Aust. J. Zool.* 15:7–27.

Burbidge, A. A., and Main, A. R. 1971. Report on a visit of inspection to Barrow Island, November, 1969. *Rep. Fish. Fauna West. Aust.* 8:1–26.

Burbidge, N. T. 1960. The phytogeography of the Australian region. *Aust. J. Bot.* 8:75–211.

Bustard, H. R. 1968a. The ecology of the Australian gecko *Gehyra variegata* in northern New South Wales. *J. Zool., Lond.* 154:113–138.

———. 1968b. The ecology of the Australian gecko *Heteronota binoei* in northern New South Wales. *J. Zool., Lond.* 156:483–497.

———. 1969. The population ecology of the gekkonid lizard *Gehyra variegata* (Dumeril and Bibron) in exploited forests in northern New South Wales. *J. Anim. Ecol.* 38:35–51.

Bustard, H. R., and Hughes, R. D. 1966. Gekkonid lizards: Average ages derived from tail-loss data. *Science, N.Y.* 153:1670–1671.

Cade, T. J., and Dybas, J. A. 1962. Water economy of the budgerygah. *Auk* 79:345–364.

Cade, T. J.; Tobin, C. A.; and Gold, A. 1965. Water economy and metabolism of two estrildine finches. *Physiol. Zoöl.* 38:9–33.

Calder, W. A. 1964. Gaseous metabolism and water relations of the zebra finch *Taenopygia castanotis*. *Physiol. Zoöl.* 37:400–413.

Charley, J. L., and Cowling, S. W. 1968. Changes in soil nutrient status resulting from overgrazing in plant communities in semi-arid areas. *Proc. ecol. Soc. Aust.* 3:28–38.

Cochrane, G. R. 1968. Fire ecology in southeastern Australian sclerophyll forests. *Proc. Ann. Tall Timbers Fire Ecol. Conf.* 8:15–40.

Cole, La M. C. 1954. Population consequences of life history phenomena. *Q. Rev. Biol.* 29:103–137.

Cookson, I. C. 1953. The identification of the sporomorph *Phyllocladites* with *Dacrydium* and its distribution in southern Tertiary deposits. *Aust. J. Bot.* 1:64–70.

Cookson, I. C., and Pike, K. M. 1953. The Tertiary occurrence and distribution of *Podocarpus* (section *Dacrycarpus*) in Australia and Tasmania. *Aust. J. Bot.* 1:71–82.

————. 1954. The fossil occurrence of *Phyllocladus* and two other podocarpaceous types in Australia. *Aust. J. Bot.* 2:60–68.

Dawson, T. J., and Brown, G. D. 1970. A comparison of the insulative and reflective properties of the fur of desert kangaroos. *Comp. Biochem. Physiol.* 37:23–38.

Dawson, T. J.; Denny, M. J. S.; and Hulbert, A. J. 1969. Thermal balance of the macropod marsupial *Macropus eugenii* Desmarest. *Comp. Biochem. Physiol.* 31:645–653.

Dawson, W. R., and Bennet, A. F. 1971. Thermoregulation in the marsupial *Lagorchestes conspicillatus*. *J. Physiol., Paris* 63:239–241.

Dawson, W. R., and Fisher, C. D. 1969. Responses to temperature by the spotted nightjar (*Eurostopodus guttatus*). *Condor* 71:49–53.

Exon, N. R.; Langford-Smith, T.; and McDougall, I. 1970. The age and geomorphic correlations of deep-weathering profiles, silcrete, and basalt in the Roma-Amby Region Queensland. *J. geol. Soc. Aust.* 17:21–31.

Fisher, C. D.; Lindgren, E.; and Dawson, W. R. 1972. Drinking patterns and behaviour of Australian desert birds in relation to their ecology and abundance. *Condor* 74:111–136.

Frakes, L. A., and Kemp, E. M. 1972. Influence of continental positions on early Tertiary climates. *Nature, Lond.* 240:97–100.

Gardner, C. A. 1944. Presidential address: The vegetation of Western Australia. *J. Proc. R. Soc. West. Aust.* 28:xi–lxxxvii.

———. 1957. The fire factor in relation to the vegetation of Western Australia. *West. Aust. Nat.* 5:166–173.

Gilbert, J. M. 1958. Forest succession in the Florentine Valley, Tasmania. *Pap. Proc. R. Soc. Tasm.* 93:129–151.

Greenwald, L.; Stone, W. B.; and Cade, T. J. 1967. Physiological adjustments of the budgerygah (*Melopsettacus undulatus*) to dehydrating conditions. *Comp. Biochem. Physiol.* 22:91–100.

Heirtzler, J. R.; Dickson, G. O.; Herron, E. M.; Pitman, W. C.; and Le Pichon, X. 1968. Marine magnetic anomalies, geomagnetic field reversals, and motions of the ocean floor and continents. *J. geophys. Res.* 73:2119–2136.

Jackson, W. D. 1965. Vegetation. In *Atlas of Tasmania*, ed. J. L. Davis, pp. 50–55. Hobart, Tasm.: Mercury Press.

———. 1968*a*. Fire and the Tasmanian flora. In *Tasmanian year book no. 2*, ed. R. Lakin and W. E. Kellend. Hobart, Tasm.: Commonwealth Bureau of Census and Statistics, Hobart Branch.

———. 1968*b*. Fire, air, water and earth: An elemental ecology of Tasmania. *Proc. ecol. Soc. Aust.* 3:9–16.

Johnson, O. W., and Mugaas, J. N. 1970. Quantitative and organizational features of the avian renal medulla. *Condor* 72:288–292.

Keast, A. 1959. Australian birds. Their zoogeography and adaptation to an arid continent. In *Biogeography and ecology in Australia*, ed. A. Keast, R. L. Crocker, and C. S. Christian, pp. 89–114. The Hague: Dr. W. Junk.

King, C. E., and Anderson, W. W. 1971. Age specific selection, II. The interaction between r & K during population growth. *Am. Nat.* 105:137–156.

Kirkpatrick, T. H. 1965. Studies of Macropodidae in Queensland. 2. Age estimation in the grey kangaroo, the eastern wallaroo and the red-necked wallaby, with notes on dental abnormalities. *Qd J. agric. Anim. Sci.* 22:301–317.

Le Pichon, X. 1968. Sea-floor spreading and continental drift. *J. geophys. Res.* 73:3661–3697.

Licht, P.; Dawson, W. R.; and Shoemaker, V. H. 1966*a*. Heat resistance of some Australian lizards. *Copeia* 1966:162–169.

Licht, P.; Dawson, W. R.; Shoemaker, V. H.; and Main, A. R. 1966*b*. Observations on the thermal relations of Western Australian lizards. *Copeia* 1966:97–110.

MacArthur, R. H., and Wilson, E. O. 1967. *The theory of island biogeography.* Monographs in Population Biology, 1. Princeton: Princeton Univ. Press.

MacMillan, R. E., and Lee, A. K. 1967. Australian desert mice: Independence of exogenous water. *Science, N.Y.* 158:383–385.

————. 1969. Water metabolism of Australian hopping mice. *Comp. Biochem. Physiol.* 28:493–514.

————. 1970. Energy metabolism and pulmocutaneous water loss of Australian hopping mice. *Comp. Biochem. Physiol.* 35:355–369.

McWhae, J. R. H.; Playford, P. E.; Lindner, A. W.; Glenister, B. F.; and Balme, B. E. 1958. The stratigraphy of Western Australia. *J. geol. Soc. Aust.* 4:1–161.

Main, A. R. 1971. Measures of well-being in populations of herbivorous macropod marsupials. In *Dynamics of populations*, ed. P. J. den Boer and G. R. Gradwell, pp. 159–173. Wageningen: PUDOC.

Mertz, D. B. 1971. Life history phenomena in increasing and decreasing population. In *Statistical ecology, volume II: Sampling and modeling biological populations and population dynamics*, ed. G. P. Patil, E. C. Pielou, and W. E. Waters, pp. 361–399. University Park: Pa. St. Univ. Press.

Moir, R. J.; Somers, M.; and Waring, H. 1956. Studies in marsupial nutrition: Ruminant-like digestion of the herbivorous marsupial *Setonix brachyurus* (Quoy and Gaimard). *Aust. J. biol. Sci.* 9:293–304.

Mount, A. B. 1964. The interdependence of eucalypts and forest fires in southern Australia. *Aust. For.* 28:166–172.

————. 1969. Eucalypt ecology as related to fire. *Proc. Ann. Tall Timbers Fire Ecol. Conf.* 9:75–108.

Murphy, G. I. 1968. Pattern in life history and the environment. *Am. Nat.* 102:391–404.

Oksche, A.; Farner, D. C.; Serventy, D. L.; Wolff, F.; and Nicholls,

C. A. 1963. The hypothalamo-hypophysial neurosecretory system of the zebra finch, *Taeniopygia castanotis. Z. Zellforsch. mikrosk. Anat.* 58:846–914.

Packham, G. H., ed. 1969. The geology of New South Wales. *J. geol. Soc. Aust.* 16:1–654.

Pianka, E. R. 1969. Sympatry of desert lizards (*Ctenotus*) in Western Australia. *Ecology* 50:1012–1030.

———. 1970. On r and K selection. *Am. Nat.* 104:592–597.

———. 1971*a*. Comparative ecology of two lizards. *Copeia* 1971:129–138.

———. 1971*b*. Ecology of the agamid lizard *Amphibolurus isolepis* in Western Australia. *Copeia* 1971:527–536.

———. 1971*c*. Notes on the biology of *Amphibolurus cristatus* and *Amphibolurus scutulatus. West. Aust. Nat.* 12:36–41.

Pianka, E. R., and Pianka, H. D. 1970. The ecology of *Moloch horridus* (Lacertilia: Agamidae) in Western Australia. *Copeia* 1970:90–103.

Ride, W. D. L. 1964. A review of Australian fossil marsupials. *J. Proc. R. Soc. West. Aust.* 47:97–131.

———. 1971. On the fossil evidence of the evolution of the Macropodidae. *Aust. Zool.* 16:6–16.

Shield, J. W. 1965. A breeding season difference in two populations of the Australian macropod marsupial *Setonix brachyurus. J. Mammal.* 45:616–625.

Simpson, G. G. 1961. Historical zoogeography of Australian mammals. *Evolution* 15:431–446.

Slatyer, R. O., and Perry, R. A., eds. 1969. *Arid lands of Australia.* Canberra: Aust. Nat. Univ. Press.

Sporn, G. G. 1955. The breeding of the mountain devil in captivity. *West. Aust. Nat.* 5:1–5.

———. 1958. Further observations on the mountain devil in captivity. *West. Aust. Nat.* 6:136–137.

———. 1965. Additional observations on the life history of the mountain devil (*Moloch horridus*) in captivity. *West. Aust. Nat.* 9:157–159.

Stirton, R. A.; Tedford, R. D.; and Miller, A. H. 1961. Cenozoic stratigraphy and vertebrate palaeontology of the Tirari Desert, South Australia. *Rep. S. Aust. Mus.* 14:19–61.

Stirton, R. A.; Tedford, R. H.; and Woodburne, M. O. 1968. Australian Tertiary deposits containing terrestrial mammals. *Univ. Calif. Publs geol. Sci.* 77:1–30.

Stirton, R. A.; Woodburne, M. O.; and Plane, M. D. 1967. A phylogeny of Diprotodontidae and its significance in correlation. *Bull. Bur. Miner. Resour. Geol. Geophys. Aust.* 85:149–160.

Storr, G. M. 1967. Geographic races of the agamid lizard *Amphibolurus caudicinctus*. *J. Proc. R. Soc. West. Aust.* 50:49–56.

———. 1968. Diet of kangaroos (*Megaleia rufa* and *Macropus robustus*) and merino sheep near Port Hedland, Western Australia. *J. Proc. R. Soc. West. Aust.* 51:25–32.

Storr, G. M.; Green, J. W.; and Churchill, D. M. 1959. The vegetation of Rottnest Island. *J. Proc. R. Soc. West. Aust.* 42:70–71.

Veevers, J. J. 1967. The Phanerozoic geological history of northwest Australia. *J. geol. Soc. Aust.* 14:253–271.

———. 1971. Phanerozoic history of Western Australia related to continental drift. *J. geol. Soc. Aust.* 18:87–96.

Vine, F. J. 1966. Spreading of the ocean floor: New evidence. *Science, N.Y.* 154:1405–1415.

———. 1970. Ocean floor spreading. *Rep. Aust. Acad. Sci.* 12:7–24.

Wallace, W. R. 1966. Fire in the Jarrah forest environment. *J. Proc. R. Soc. West. Aust.* 49:33–44.

Watkins, J. R. 1967. The relationship between climate and the development of landforms in the Cainozoic rocks of Queensland. *J. geol. Soc. Aust.* 14:153–168.

Wild, A. 1958. The phosphate content of Australian soils. *Aust. J. agric. Res.* 9:193–204.

Winkworth, R. E. 1967. The composition of several arid spinifex grasslands of central Australia in relation to rainfall, soil water relations, and nutrients. *Aust. J. Bot.* 15:107–130.

Woodburne, M. O. 1967. Three new diprotodontids from the Tertiary of the Northern Territory. *Bull. Bur. Miner. Resour. Geol. Geophys. Aust.* 85:53–104.

Woolnough, W. G. 1928. The chemical criteria of peneplanation. *J. Proc. R. Soc. N.S.W.* 61:17–53.

6. Species and Guild Similarity of North American Desert Mammal Faunas: A Functional Analysis of Communities   James A. MacMahon

Introduction

A major thrust of current ecological and evolutionary research is the analysis of patterns of species diversity or density in similar or vastly dissimilar community types. Such studies are believed to bear on questions concerned with the nature of communities and their stability (e.g., MacArthur, 1972), the concept of ecological equivalents or ecospecies (Odum, 1969 and 1971; Emlen, 1973), and, of course, the nature of the "niche" (Whittaker, Levin, and Root, 1973).

An approach emerging from this plethora is that of functional analysis of community components: attempts to compare the functionally similar community members, regardless of their taxonomic affinities. Root (1967, p. 335) coined the term *guild* to define "a group of species that exploit the same class of environmental resources in a similar way. This term groups together species without regard to taxonomic position, that overlap significantly in their niche requirements."

Guild is clearly differentiated from *niche* and *ecotope*, recently redefined and defined respectively (Whittaker et al., 1973, p. 335) as "applying 'niche' to the role of the species within the community, 'habitat' to its distributional response to intercommunity environmental factors, and 'ecotope' to its full range of adaptations to external factors of both niche and habitat." *Guild* groups parts of species' niches permitting intercommunity comparisons.

Without referring to the semantic problems, Baker (1971) used such a "functional" approach when he compared nutritional strategies of North American grassland myomorph rodents. Wiens (1973) developed a similar theme in his recent analysis of grassland bird communities, as did Wilson (1973) with an analysis of bat faunas, and Brown (1973) with rodents of sand-dune habitats.

This paper is an attempt to compare species and functional analyses of the small mammal component of North American deserts and to use these analyses to discuss some aspects of the broader ecological and evolutionary questions of "similarity" and function of communities.

Sites and Techniques

Sites

The data base for this study is simply the species lists for a number of desert or semidesert grassland localities in the western United States. The list for a locality represents those species that occur on a piece of landscape of 100 ha in extent. This size unit allows the inclusion of spatial heterogeneity.

The localities used, the data source, and the abbreviations to be used subsequently are Jornada Bajada (*j*), a Chihuahuan desert shrub community near Las Cruces, New Mexico, operated by New Mexico State University as part of the US/IBP Desert Biome studies; Jornada Playa (*jp*), a desert grassland and mesquite area a few meters from *j* operated under the same program; Portal, Arizona (*cc*), a semidesert scrub area studied extensively by Chew and Chew (1965, 1970); Santa Rita Experimental Range (*sr*), south of Tucson, Arizona, an altered desert grassland studied by University of Arizona personnel for the US/IBP Desert Biome; Tucson Silverbell Bajada (*t*), a typical Sonoran desert (*Larrea Cercidium*) community northwest of Tucson, Arizona, operated as *sr*; Big Bend National Park (*bb*), a Chihuahuan desert shrub community near the park headquarters typified by Denyes (1956) and K. L. Dixon (1974, personal communication); Deep Canyon, California (*dc*), studied by Ryan (1968) and Joshua Tree National Monument (*jt*) studied by Miller and Stebbins (1964)—both *Larrea*-dominated areas in a transition from a Sonoran desert subdivision (Coloradan) but including many Mojave desert elements; Rock Valley (*rv*), northwest of Las Vegas, Nevada, on the

Fig. 6-1. The location of sites discussed (abbreviations explained in text) and outline of mammal provinces adapted from Hagmeier and Stults (1964): I. Artemisian; II. Mojavian; III. Navajonian; IV. Sonoran; V. Yaquinian; VI. Mapimi.

Atomic Energy Commission's Nevada Test Site, a Mojave desert shrub site operated as part of the US/IBP Desert Biome by personnel of the Environmental Biology Division of the Laboratory of Nuclear Medicine and Radiation Biology of the University of California, Los Angeles; and Curlew Valley (c), a Great Basin desert, sagebrush site of the US/IBP Desert Biome operated by Utah State University. The positions of the sites are summarized in figure 6-1. All sites have been visited and observed by me.

Analyses

Similarity was calculated using a modified form of Jaccard analysis (community coefficients) (Oosting, 1956; see also MacMahon and Trigg, 1972):

$$\frac{2w}{a + b} \times 100$$

where w is the number of species common to both faunas being compared, a is the number of species in the smaller fauna, and b is the number of species in the larger fauna.

Species similarity merely uses different taxa as units for calculations. Functional similarity uses functional units (guilds) based mainly on food habits and adult size of nonflying mammals, jack rabbit in size or smaller. The twelve desert guilds recognized, with examples of species from a single locality, include five granivores (two possible dormant-season divisions) (*Dipodomys spectabilis*, *D. merriami*, *Perognathus penicillatus*, *P. baileyi*, *P. amplus*); a "carnivorous" mouse (*Onychomys torridus*); a large and small browser (*Lepus californicus*, *Sylvilagus auduboni*); two micro-omnivores (*Peromyscus eremicus*, *P. maniculatus*); a "pack rat" (*Neotoma albigula*); and a diurnal medium-sized omnivore (*Citellus tereticaudus*). When grassland guilds are mentioned, two grazers are added to the above. Data for all pair-wise comparisons of sites are summarized in figure 6-2.

The list of mammals for all sites includes forty-seven species in fifteen genera. An additional fifteen or so species occur near the sites but were not collected on the prescribed areas.

SPECIES

	j	jp	cc	sr	t	bb	dc	jt	rv	c
j	■	67	67	41	41	56	33	30	20	14
jp	67	■	63	52	39	46	32	23	19	19
cc	79	63	■	60	56	54	32	29	19	14
sr	63	60	52	■	63	38	29	27	15	25
t	85	79	56	55	■	48	33	30	20	04
bb	80	62	85	69	85	■	40	44	48	08
dc	85	67	67	63	71	80	■	63	60	09
jt	86	69	69	65	73	89	73	■	73	08
rv	85	67	67	55	85	80	85	86	■	14
c	60	67	56	55	50	72	60	53	71	■

F
U
N
C
T
I
O
N
S

Fig. 6-2. Similarity analysis (%) matrix derived from Jaccard analysis (see text): species comparisons above the diagonal, guild comparisons below the diagonal.

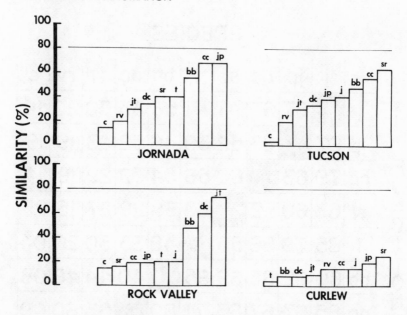

Fig. 6-3. Comparison of the similarity (%) of lists of nonflying mammal species on all sites to those of four "typical" North American desert sites: Jornada (j), Chihuahuan desert; Tucson (t), Sonoran desert; Rock Valley (rv), Mojave Desert; Curlew (c), Great Basin (cold desert).

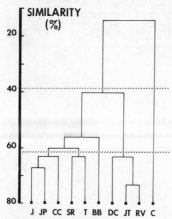

Fig. 6-4. Dendrogram showing relationships between species composition (maximum percent similarity, using Jaccard analysis) at North American sites (see text for abbreviations). The levels of significance for provinces (about 62%) and for superprovinces (about 39%), as defined by Hagmeier and Stults (1964), are marked.

Results and Discussion

Species Density

Figure 6-3 depicts the comparison of the species composition of all sites with that of each of the four "typical" desert sites of the US/IBP Desert Biome. These sites represent each of the four North American deserts: three "hot" deserts—Chihuahuan (j), Sonoran (t), Mojave (rv); one "cold" desert—Great Basin (c). Sites jp, sr, and cc are considered to have strong desert grassland affinities; all were "good" grasslands in historical times (Gardner, 1951; Lowe, 1964; Lowe et al., 1970). It is clear that none of the comparisons indicates high similarity (operationally defined as 80%). Comparison of figure 6-3 and figure 6-1 indicates that what similarity exists seems to be due to geographic proximity.

Maximum Species Similarity

The maximum species similarity of all sites is used to develop a dendrogram of relationships of sites similar to those of Hagmeier and Stults (1964) (fig. 6-4). The five groupings derived (using a 60% similarity level as was used by Hagmeier and Stults)—that is, j, jp, cc; sr, t; bb; dc, jt, rv; and c—do not follow closely the mammal provinces erected by Hagmeier and Stults (1964) and are redrawn here (fig. 6-1).

There is agreement between my data and those of Hagmeier and Stults at the superprovince level (about 38% similarity) which sets c (Artemisian) apart from all hot desert sites. The Artemisian province is equivalent to the Great Basin desert or cold desert in extent. The failure of this study and that of Hagmeier and Stults to agree may be due to the animal size limit used herein (smaller than jack rabbits) or the confined areal sample (100 ha vs. larger areas).

The groups defined here (at the province level) do seem to have some faunal meaning: there is a distinct Big Bend (bb) assemblage, a Sonoran desert group (sr, t), a Chihuahuan desert group (j, jp, cc), a Mojave desert group (dc, jt, rv), and a Great Basin desert group (c).

Some groups can be explained utilizing the evolution-biogeography discussion of Findley (1969) which differentiated eastern and western

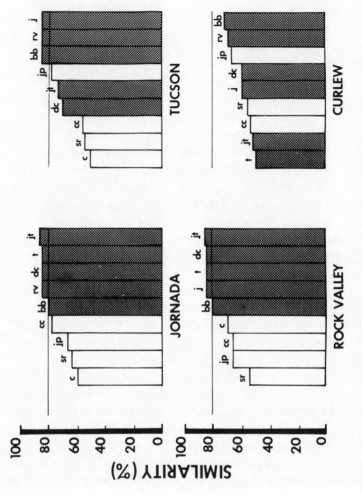

Fig. 6-5. Comparison of similarity in guilds of nonflying mammals at all sites. Presentation as in figure 6-3, except that the bars of all "hot" desert localities are shaded.

desert components meeting in southeastern Arizona—this coincides with the Sonoran desert versus the Chihuahuan desert above, with the *cc* site being intermediate. Figure 6-2 supports the intermediacy of *cc*. Findley postulated that the Deming Plain was a barrier to desert mammal movement in pluvial times, that it limited gene flow, and that it permitted speciation to the west and east. There is sharp differentiation of these two components from the Mojave and Great Basin, which is expected on the basis of their different geological histories, and also from the Big Bend, which might be more representative of the major portion of the Chihuahuan desert mammal fauna in Mexico.

Functional Diversity

When the forty-seven species of mammals are placed in guilds, rather than being treated as taxa, there is a high degree of similarity among all hot desert sites, but significant differences persist between the cold desert (*c*) and hot deserts (fig. 6-5).

A further indication of the biological soundness of guilds follows from a comparison of each desert with its geographically closest desert or destroyed grassland (*jp*, *cc*, or *sr*). This comparison generally indicates no increase in similarity whether using taxa or guilds (table 6-1). If some distinctly grassland guilds (grazers) are added

Table 6-1. *Similarity between Desert Grasslands and Closest Deserts*

Sites Compared	Similarity Index By Species	By Guilds
j-jp	67	67
j-cc	67	79
t-cc	56	56
t-sr	63	55

Note: Figures represent percent similarity coefficient (Jaccard analysis) of each desert (altered) grassland with its geographically closest desert on the basis of both species composition of fauna and functional groups (guilds).

(table 6-2), similarity of grasslands to each other rises from low levels to those considered significant. Nonsimilarity was then a problem of not including enough specifically grassland guilds.

Comparison of levels of similarity is significant and explains the operational definition of adequate similarity. Using mean similarity values (mean ± standard error) calculated from data in figure 6-2, functional (guild) similarity among hot deserts is 81.87 ± 1.43 percent; grassland to other grasslands, 87.33 ± 0.33 percent; hot deserts to grasslands, 66.32 ± 1.88 percent; cold desert to hot desert, 61.0 ± 3.69 percent; and cold desert to grassland, 59.33 ± 3.85 percent. Three functional categories are clear: hot desert, cold desert, and grassland. Cross comparison of the means of functional categories versus nonfunctional ones demonstrates significant (.001 level) differences. The Behrens-Fisher modification of the t test and Cochran's approximation of t' were used because of the heterogeneity of sample variances (Snedecor and Cochran, 1967).

Table 6-2. *Similarity among Altered Grassland Sites*

| Sites Compared | Similarity Indexes | | |
	By Species	By Guild	By Guilds (including two grassland guilds)
jp-cc	63	63	87
cc-sr	60	52	88
sr-jp	52	60	87

Note: Figures represent percent similarity coefficient (Jaccard analysis) between destroyed or altered grassland sites, using species composition and functional units (guilds) and adding two specifically grassland guilds.

General Discussion

Causes of Species Mixes

An implication of the data presented here is that any one of a number of species of small mammals may be functionally similar in a particular example of desert scrub community. It is well known that various species of mice in the genera *Microdipidops*, *Peromyscus*, and *Perognathus* may have overlapping ranges and be desert adapted, but seem to replace each other in specific habitats of various soil-texture characteristics (sandy to rocky) (Hardy, 1945; Ryan, 1968; Ghiselin, 1970). Soil surface strength, and vegetation height and density, explain relative densities of some desert small mammals (Rosenzweig and Winakur, 1969; Rosenzweig, 1973). Interspecific behavior is part of the partitioning of habitats by a number of desert rodents: *Neotoma* (Cameron, 1971); *Dipodomys* (Christopher, 1973); and a seven-species "community" (MacMillan, 1964). Brown (1973) and Brown and Lieberman (1973) attribute species diversity of sand-dune rodents to a mix of ecological, biogeographic, and evolutionary factors, a position similar to that I take for a broader range of desert conditions.

Many other factors are also involved in defining various axes of specific niches (*sensu* Whittaker et al., 1973) of desert mammals; these specific niche differences do not exclude the functional overlap of other niche components. These examples and others merely show how finely genetically plastic organisms can subdivide the environment. These differences need not change the basic role of the species.

If the important shrub community function is seed removal, it does not matter to the community what species does it. The removal could be by any one of several rodents differing in soil-texture preferences or perhaps by a bird or even ants. Ecological equivalents may or may not be genetically close. All niche axes of a species or population are not equally important to the functioning of the community.

Importance of Functional Approach

Since all three hot deserts have similar functional diversity but vary in basic ways (e.g., more rain, more vegetational synusia, biseasonal rainfall pattern in the Sonoran desert as compared to the Mojave), functional diversity does not correlate well with those factors thought to relate animal species diversity to community structure—that is, vertical and horizontal foliage complexity (Pianka, 1967; MacArthur, 1972).

The crux of the problem is that most measures of species diversity include some measure of abundance (Auclair and Goff, 1971). The analysis herein counts only presence or absence—a level of abstraction more general than species diversity measures.

This greater generality is justified, I believe, because it strikes closer to the problem of comparing community types which are basically similar (e.g., hot deserts) but include units each having undergone specific development in time and space (e.g., Mojave vs. Sonoran vs. Chihuahuan hot deserts). The analysis seeks to elucidate various levels of the least-common denominators of community function.

The guild level of abstraction has potential applied value. The generalization of forty-seven mammal species into only twelve functional groups may permit the development of suitable predictive computer models of "North American Hot Deserts." The process of accounting for the vagaries of every species makes the task of modeling cumbersome and may prohibit rapid expansion of this ecological tool. The abstraction of a large number of species into biologically determined "black boxes" is an acceptable compromise.

There is clear evidence that temporal variations in the density of mammal species may so affect calculations of species diversity that their use and interpretation in a community context are difficult (M'Closkey, 1972). While such variations are intrinsically interesting biologically, they should not prevent us from seeking more generally applicable, albeit less detailed, predictors of community organization.

Guilds, Niches, Species, Stability

The guilds chosen here were selected on the basis of subjective familiarity with desert mammals. These guilds may not be requisites for the community as a whole to operate. If one were able to perceive the *requisite* guilds of a community, several things seem reasonable. First, to be stable (able to withstand perturbations without changing basic structure and function) a community cannot lose a requisite guild. Species performing functions provide a stable milieu if, first, they are themselves resilient to a wide range of perturbations (i.e., no matter what happens they survive) or, second, each requisite function can be performed by any one of a number of species, despite niche differences among these species—that is, the community contains a high degree of functional redundancy, preventing or reducing changes in community characteristics. The importance of this was alluded to by Whittaker and Woodwell (1972), who cited the case of oaks replacing chestnuts after they were wiped out by the blight in the North American eastern deciduous forests, and on theoretical grounds by Conrad (1972).

Any stable community is some characteristic mix of resilient and redundant species. Species diversity per se may not then correlate well with stability. Stability might come with a number of species diversities as long as the requisite guilds are represented. Tropics and deserts might represent extremes of a large number of redundant species as opposed to fewer more resilient species, both mixes conferring some level of stability as witnessed by the historical persistence of these community types.

None of this implies that all species are requisite to a community, or that coevolution is the only path for community evolution. Many species may be "tolerated" by communities just because the community is well enough buffered that minor species have no noticeable effect. As long as they pass their genetic make-up on to a new generation, species are successful; they need not do anything for the community.

Acknowledgments

These studies were made possible by a National Science Foundation grant (GB 32139) and are part of the contributions of the US/IBP Desert Biome Program. I am indebted to the following people for data collected by them or under their supervision: W. Whitford, E. L. Cockrum, K. L. Dixon, F. B. Turner, B. Maza, R. Anderson, and D. Balph. F. B. Turner and N. R. French kindly commented on a version of the manuscript.

Summary

The nonflying, small mammal faunas of western United States deserts were compared (coefficient of community) on the basis of their species and guild (functional) composition. Guilds were derived from information on animal size and food habits.

It is concluded that the similarity among sites with respect to guilds, though the species may differ, is a result of a complex of evolutionary events and particular contemporary community characteristics of the specific sites. Functional similarity, based on functional groups (guilds), is rather constant among hot deserts and different between hot deserts and either cold desert or desert grassland.

The functional analysis describes only a part of the niche of an organism, but perhaps an important part. Such abstractions and generalizations of the details of the community's complexities permit mathematical modeling to progress more rapidly and allow address to the general question of community "principles."

Guilds required by communities to maintain community integrity against perturbations may be better correlates to community stability than the various measures of species diversity currently popular.

References

Auclair, A. N., and Goff, F. G. 1971. Diversity relations of upland forests in the western Great Lakes area. *Am. Nat.* 105:499–528.

Baker, R. H. 1971. Nutritional strategies of myomorph rodents in North American grasslands. *J. Mammal.* 52:800–805.

Brown, J. H. 1973. Species diversity of seed-eating desert rodents in sand dune habitats. *Ecology* 54:775–787.

Brown, J. H., and Lieberman, G. A. 1973. Resource utilization and co-existence of seed-eating desert rodents in sand dune habitats. *Ecology* 54:788–797.

Cameron, G. N. 1971. Niche overlap and competition in woodrats. *J. Mammal.* 52:288–296.

Chew, R. M., and Chew, A. E. 1965. The primary productivity of a desert shrub (*Larrea tridentata*) community. *Ecol. Monogr.* 35:355–375.

———. 1970. Energy relationships of the mammals of a desert shrub *Larrea tridentata* community. *Ecol. Monogr.* 40:1–21.

Christopher, E. A. 1973. Sympatric relationships of the kangaroo rats, *Dipodomys merriami* and *Dipodomys agilis*. *J. Mammal.* 54:317–326.

Conrad, M. 1972. Stability of foodwebs and its relation to species diversity. *J. theoret. Biol.* 32:325–335.

Denyes, H. A. 1956. Natural terrestrial communities of Brewster County, Texas, with special reference to the distribution of mammals. *Am. Midl. Nat.* 55:289–320.

Emlen, J. M. 1973. *Ecology: An evolutionary approach.* Reading, Mass: Addison-Wesley.

Findley, J. S. 1969. Biogeography of southwestern boreal and desert mammals. *Univ. Kans. Publs Mus. nat. Hist.* 51:113–128.

Gardner, J. L. 1951. Vegetation of the creosotebush area of the Rio Grande Valley in New Mexico. *Ecol. Monogr.* 21:379–403.

Ghiselin, J. 1970. Edaphic control of habitat selection by kangaroo mice (*Microdipodops*) in three Nevadan populations. *Oecologia* 4:248–261.

Hagmeier, E. M., and Stults, C. D. 1964. A numerical analysis of the distributional patterns of North American mammals. *Syst. Zool.* 13:125–155.

Hardy, R. 1945. The influence of types of soil upon the local distribution of some small mammals in southwestern Utah. *Ecol. Monogr.* 15:71–108.

Lowe, C. H. 1964. Arizona landscapes and habitats. In *The vertebrates of Arizona*, ed. C. H. Lowe, pp. 1–132. Tucson: Univ. of Ariz. Press.

Lowe, C. H.; Wright, J. W.; Cole, C. J.; and Bezy, R. L. 1970. Natural hybridization between the teiid lizards *Cnemidophorus sonorae* (parthenogenetic) and *Cnemidophorus tigris* (bisexual). *Syst. Zool.* 19:114–127.

MacArthur, R. H. 1972. *Geographical ecology.* New York: Harper & Row.

M'Closkey, R. T. 1972. Temporal changes in populations and species diversity in a California rodent community. *J. Mammal.* 53:657–676.

MacMahon, J. A., and Trigg, J. R. 1972. Seasonal changes in an old-field spider community with comments on techniques for evaluating zoosociological importance. *Am. Midl. Nat.* 87:122–132.

MacMillan, R. E. 1964. Population ecology, water relations and social behavior of a southern California semidesert rodent fauna. *Univ. Calif. Publs Zool.* 71:1–66.

Miller, A. H., and Stebbins, R. C. 1964. *The lives of desert animals in Joshua Tree National Monument.* Berkeley and Los Angeles: Univ. of Calif. Press.

Odum, E. P. 1969. The strategy of ecosystem development. *Science, N.Y.* 164:262–270.

———. 1971. *Fundamentals of ecology*. 3d ed. Philadelphia: W. B. Saunders Co.

Oosting, H. J. 1956. *The study of plant communities*. 2d ed. San Francisco: Freeman Co.

Pianka, E. R. 1967. On lizard species diversity: North American flatland deserts. *Ecology* 48:333–351.

Root, R. B. 1967. The niche exploitation pattern of the blue-gray gnatcatcher. *Ecol. Monogr.* 37:317–350.

Rosenzweig, M. L. 1973. Habitat selection experiments with a pair of co-existing heteromyid rodent species. *Ecology* 54:111–117.

Rosenzweig, M. L., and Winakur, J. 1969. Population ecology of desert rodent communities: Habitats and environmental complexity. *Ecology* 50:558–572.

Ryan, R. M. 1968. *Mammals of Deep Canyon*. Palm Springs, Calif.: Desert Museum.

Snedecor, G. W., and Cochran, W. G. 1967. *Statistical methods*. 6th ed. Ames: Iowa State Univ. Press.

Whittaker, R. H., and Woodwell, G. M. 1972. Evolution of natural communities. In *Ecosystem structure and function*, ed. J. Wiens, pp. 137–156. Corvallis: Oregon State Univ. Press.

Whittaker, R. H.; Levin, S. A.; and Root, R. B. 1973. Niche, habitat, and ecotope. *Am. Nat.* 107:321–338.

Wiens, J. A. 1973. Pattern and process in grassland bird communities. *Ecol. Monogr.* 43:237–270.

Wilson, D. E. 1973. Bat faunas: A trophic comparison. *Syst. Zool.* 22.14–29.

7. The Evolution of Amphibian Life Histories in the Desert Bobbi S. Low

Introduction

Among desert animals amphibians are especially intriguing because at first glance they seem so obviously unsuited to arid environments. Most amphibians require the presence of free water at some stage in the life cycle. Their skin is moist and water permeable, and their eggs are not protected from water loss by any sort of tough shell. Perhaps the low number of amphibian species that live in arid regions reflects this.

Some idea of the variation in life-history patterns which succeed in arid regions is necessary before examining the environmental parameters which shape those life histories. Consider three different strategies. Members of the genus *Scaphiopus* found in the southwestern United States frequent short-grass plains and alkali flats in arid and semiarid regions and are absent from high mountain elevations and extreme deserts (Stebbins, 1951). Species of the genus breed in temporary ponds and roadside ditches, often on the first night after heavy rains. *Scaphiopus bombifrons* lays 10 to 250 eggs in a number of small clusters; *Scaphiopus hammondi* lays 300 to 500 eggs with a mean of 24 eggs per cluster; and *S. couchi* lays 350 to 500 eggs in a number of small clusters. All three species burrow during dry periods of the year.

The genus *Bufo* is widespread, and a large number of species live in arid and semiarid regions. *Bufo alvarius* lives in arid regions but, unlike *Scaphiopus*, appears to be dependent on permanent water. Stebbins (1951) notes that, while summer rains seem to start seasonal activity, such rains are not always responsible for this activity.

The mating call has been lost. Also, unlike *Scaphiopus*, this toad lays between 7,500 and 8,000 eggs in one place at one time.

Eleutherodactylus latrans occurs in arid and semiarid regions and does not require permanent water. It frequents rocky areas and canyons and may be found in crevices, caves, and even chinks in stone walls. In Texas, Stebbins (1951) reported that *E. latrans* becomes active during rainy periods from February to May. About fifty eggs are laid on land in seeps, damp places, or caves; and, unlike either *Scaphiopus* or *Bufo*, the male may guard the eggs.

These three life histories diverge in degree of dependence on water, speed of breeding response, degree of iteroparity, and amount and kind of reproductive effort per offspring or parental investment (Trivers, 1972). All three strategies may have evolved, and at least are successful, in arid regions.

Two forces will shape desert life histories. The first is the relatively high likelihood of mortality as a result of physical extremes. Considerable work has been done on the mechanics of survival in amphibians which live in arid situations (reviewed by Mayhew, 1968). Most of the research concentrated on physiological parameters like dehydration tolerance, ability to rehydrate from damp soil, and speed of rehydration (Bentley, Lee, and Main, 1958; Main and Bentley, 1964; Warburg, 1965; Dole, 1967); water retention and cocoon formation (Ruibal, 1962a, 1962b; Lee and Mercer, 1967); and the temperature tolerances of adults and tadpoles (Volpe, 1953; Brattstrom, 1962, 1963; Heatwole, Blasina de Austin, and Herrero, 1968). Bentley (1966), reviewing adaptations of desert amphibians, gave the following list of characteristics important to desert species:

1. No definite breeding season
2. Use of temporary water for reproduction
3. Initiation of breeding behavior by rainfall
4. Loud voices in males, with marked attraction of both males and females, and the quick building of large choruses
5. Rapid egg and larval development
6. Ability of tadpoles to consume both animal and vegetable matter
7. Tadpole cannibalism
8. Production of growth inhibitors by tadpoles

9. High heat tolerance by tadpoles
10. Metatarsal spade for burrowing
11. Dehydration tolerance
12. Nocturnal activity

Bentley's list consists mostly of physiological or anatomical characteristics associated with survival in the narrow sense. Only two or three items involve special aspects of life cycles, and some characteristics as stated are not exclusive to desert forms. Most investigators have emphasized the problems of survival for desert amphibians for the obvious reason that the animal and its environment seem so ill-matched, and most investigators have emphasized morphological and physiological attributes because they are easier to measure. A notable exception to this principally anatomical or physiological approach is the work of Main and his colleagues (Main, 1957, 1962, 1965, 1968; Main, Lee, and Littlejohn, 1958; Main, Littlejohn, and Lee, 1959; Lee, 1967; Martin, 1967; Littlejohn, 1967, 1971) who have discussed life-history adaptations of Australian desert anurans. Main (1968) has summarized some general life-history phenomena that he considered important to arid-land amphibians including high fecundity, short larval life, and burrowing. However, as he implied, the picture is not simple. A surprising variety of successful life-history strategies exists in arid and semiarid amphibian species, far greater than one would predict from attempts (Bentley, 1966; Mayhew, 1968) to summarize desert adaptations in amphibians. If survival were the critical focus of selection, one might predict fewer successful strategies and more uniformity in the kinds of life histories successful in arid-land amphibians.

But succeeding in the desert, as elsewhere, is a matter of balancing risk of mortality against optimization of reproductive effort so that realized reproduction is optimized. As soon as survival from generation to generation occurs, selection is then working on differences in reproduction among the survivors, an important point emphasized by Williams (1966a) in arguing that adaptations should most often be viewed as the outcomes of better-versus-worse alternatives rather than as necessities in any given circumstance. The focus I wish to develop here is on the critical parameters shaping the evolution of life-history strategies and the better-versus-worse alternatives in each of a number of situations. Adaptations of desert amphibians have scarcely been examined in this light.

Life-History Components and Environmental Parameters

Wilbur, Tinkle, and Collins (1974), in an excellent paper on the evolution of life-history strategies, list eight components of life histories: juvenile and adult mortality schedules, age at first reproduction, reproductive life span, fecundity, fertility, fecundity-age regression, degree of parental care, and reproductive effort. The last two are included in Trivers's (1972) concept of parental investment. For very few, if any, anurans are all these parameters documented.

I will concentrate here on problems of parental investment, facultative versus nonfacultative responses, cryptic versus clumping responses, and shifts in life-history stages. How does natural selection act on these traits in different environments? What environmental parameters are actually significant?

Classifying ranges of environmental variation may seem at first like a job for geographers; but, even when one acknowledges that deserts may be hot or cold, seasonal or nonseasonal, that they may possess temporary or permanent waters, different vegetation, and different soils, I think a few parameters can be shown to have overriding importance. These are (a) the *range* of the variation in the environmental attributes I have just described—temperature, humidity, day length, and so on; (b) the *predictability* of these attributes; and (c) their *distribution*—the patchiness or grain of the environment (Levins, 1968). Wilbur et al. (1974) consider trophic level and successional position also as life-history determinants in addition to environmental uncertainty. At the intraordinal level, these effects may be more difficult to sort out; and, at present, data are really lacking for anurans.

Range

Obviously the overall range of variation in environmental parameters is important in shaping patterns of behavior or life history. The same life-history strategy will not be equally successful in an environment where, for instance, temperature fluctuates only 5° daily, as in an environment in which fluctuations may be as much as 20° to 30°C. The range of fluctuations may strongly affect selection on physiological adaptations and differences in survival. Ranges of variation, particularly in temperature and water availability, are extreme in the environments of desert amphibians; but such effects have been dealt with more fully than the others I wish to discuss, and so I will concentrate on other factors.

Predictability

It is probably true that deserts are less predictable than either tropical or temperate mesic situations; Bentley's (1966) list of adaptations reflected this characteristic. The terms *uncertainty* and *predictability* are generally used for physical effects—seasonality and catastrophic events, for instance—but may include both spatial and biotic components. In fact, both patchiness and the distribution of predation mortality modify uncertainty.

Two aspects of predictability must be distinguished, for they affect the relative success of different life-history strategies quite differently. Areas may vary in reliability with regard to when or where certain events occur, such as adequate rainfall for successful breeding. Further, the suitability of such events may vary—a rain or a warm spell, whenever and wherever it may occur, may or may not be suitable for breeding. In a northern temperate environment the succession of the seasons is predictable. For a summer-breeding animal some summers will be better than others for breeding; this is reflected, for example, in Lack's (1947, 1948) results on clutch-size variation in English songbirds from year to year (see also Klomp, 1970; Hussell, 1972). Most summers, however, will be at least minimally suitable, and relatively few temperate—mesic-area organisms appear to have evolved to skip breeding in poor years. On the other hand, in most deserts rain is less predictable not only in regard to when and where it occurs, but also in regard to its effectiveness. Perhaps this latter aspect of environmental predictability has not been sufficiently emphasized in terms of its role in shaping life histories.

It is probably sufficient to distinguish four classes of environments with regard to predictability.

1. Predictable and relatively unchanging environments, such as caves and to a lesser extent tropical rain forests.
2. Predictably fluctuating or cyclic environments, areas with diurnal and seasonal periodicities, like temperate mesic areas.
3. Acyclic environments, unpredictable with reference to the timing and frequency of important events like rain, but predictable in terms of their effectiveness. If an event occurs, either it always is effective or the organism can judge the effectiveness.
4. Noncyclic environments that give few clues as to effectiveness of events: for example, rainfall erratic in spacing, timing, and amount. Areas like the central Australian desert present this situation for most frogs.

Optimal life-history strategies will differ in these environments, and desert amphibians must deal not only with extremes of temperature and aridity that seem contrary to their best interests but also with high degrees of unpredictability in those same environmental features and with localized and infrequent periods suitable for breeding.

Environmental uncertainty may have significant effects on shifts in life histories and on phenotypic similarities between life-history stages. If the duration of habitat suitable for adults is uncertain, or frequently less than one generation, the evolution of very different larval stages, not dependent on duration of the adult habitat, will be favored. The very fact that anurans show complex metamorphosis, with very different larval and adult stages, suggests this has been a factor in anuran evolution. Wilbur and Collins (1973) have discussed ecological aspects of amphibian metamorphosis and the role of uncertainty in the evolution of metamorphosis. An effect of complex metamorphosis is to increase independence of variation in the likelihoods of success in different life stages. Selective forces in the various habitats occupied by the different life stages are more likely to change independently of one another. As I will show later, this situation has profound effects on life-cycle patterns.

Predictable seasonality will favor individuals which breed seasonally during the most favorable period. Those who breed early in the good season will produce offspring with some advantage in size and feeding ability, and perhaps food availability, over the offspring of later breeders. Females which give birth or lay eggs early may, furthermore, increase their fitness and reduce their risk of feeding and improving their condition during the good season (Tinkle, 1969). Fisher (1958) has shown that theoretical equilibrium will be reached when the numbers of individuals breeding per day are normally distributed, if congenital earliness of breeding and nutritional level are also normally distributed. Predation (see below) on either eggs or breeding adults may cause amphibian breeding choruses to become clumped in space (Hamilton, 1971) and time. The timing, then, of the breeding peaks will depend on the balance between the time required after conditions become favorable for animals to attain breeding condition and the pressure to breed early. Both seasonal temperature and seasonal rainfall differences may limit breeding, and most amphibians in North American mesic areas and seasonally dry tropics (Inger and Greenberg, 1956; Schmidt and Inger, 1959) appear to breed seasonally.

In predictable unchanging environments, two strategies may be effective, depending on the presence or absence of predation. If no predation existed, individuals in "uncrowded" habitats would be selected to mature early and breed whenever they mature, maximizing egg numbers and minimizing parental investment per offspring, while individuals in habitats of high interspecific competition would be selected for the production of highly competitive offspring. That is, neither climatic change nor predation would influence selection, and MacArthur and Wilson's (1967) suggestion of r- and K- trends may hold. The result would be that adults would be found in breeding condition throughout the year. In a study by Inger and Greenberg (1963), reproductive data were taken monthly from male and female *Rana erythraea* in Sarawak. Rain and temperature were favorable for breeding throughout the year. From sperm and egg counts and assessment of secondary sex characters, they determined that varying proportions of both sexes were in breeding condition throughout the year. The proportion of breeding bore no obvious relation to climatic factors. Inger and Greenberg suggested that this situation represented the "characteristic behavior of most stock from which modern species of frogs arose." If predation exists in nonseasonal environments, year-round breeding with cryptic behavior may be successful; but if predation is erratic or predictably fluctuating (rather than constant), a "selfish herd" strategy may be favored.

Situations in which important events are unpredictable lead to other strategies. Life where the environment is unpredictable not only as to when or where events will occur but also as to whether or not they will be effective is comparable to playing roulette on a wheel weighted in an unknown fashion. Two strategies will be at a selective advantage:

1. Placing a large number of small bets will be favored, rather than placing a small number of large bets, or placing the entire bet on one spin of the wheel. In other words, in such an unpredictable situation, one expects iteroparous individuals who will lay a few eggs each time there is a rain. A corollary to this prediction is that, when juvenile mortality is unpredictable, longer adult life as well as iteroparity will be favored (cf. Murphy, 1968).

2a. Any strategy will be favored which will help an individual to judge the effectiveness of an event (i.e., to discover the weighting of the wheel). The central Australian species of *Cyclorana*— in fact, most of the Australian deep-burrowing frogs—may represent such a case. During dry periods, *Cyclorana platycephalus*, for instance, burrows three to four feet deep in clay soils. Light rains have no effect on dormant frogs even when rain occurs right in the area, since much of it runs off and does not percolate through to the level where the frogs are burrowed. Any rain reaching the frogs, we may suppose, is likely to be sufficient for tadpoles to mature and metamorphose. Thus, whatever functions (*sensu* Williams, 1966a) burrowing may serve in *Cyclorana*, one effect is that selective advantage accrues to those burrowing deeply because reproductive effort is not expended on unsuitable events.

2b. Any behavior which makes events less random, enhancing positive effects or reducing the effects of catastrophic events, will be favored. For example, parents may be favored who lay their eggs in some manner that tends to reduce the impact of flooding on their offspring, such as by laying their eggs out of the water and in rocky crevices or up on leaves or in burrows. A number of leptodactylid frogs do this (table 7-1). Obviously, such a strategy would only be favored when it had the effect of making mortality nonrandom. In deserts, where humidity is low and evapotranspiration high, it would not appear to be a particularly effective strategy; in fact only one *Eleutherodactylus* (Stebbins, 1951) and one species of *Pseudophryne* (Main, 1965) living in arid regions appear to follow strategies of hiding their eggs (table 7-1).

When the timing of events is unpredictable, but their effectiveness is not, individuals who only respond to suitable events will obviously be favored. This situation probably never exists a priori but only because organisms living in environments unpredictable both as to timing and effectiveness will evolve to respond only to suitable events, as in the burrowing *Cyclorana*. Thus, environments in the No. 4 category above will slowly be transformed into No. 3 environments by changes in the organisms inhabiting them. This emphasizes the importance of describing environments in terms of the organisms.

In the evolution of life cycles in uncertain environments, one kind of evidence of "learning the weighting of the wheel" is the capability of quickly exploiting unpredictable breeding periods—for example, ability to start a reproductive investment quickly after a desert rainfall. Another is the ability to terminate inexpensively an investment that has become futile, such as the care of offspring begun during a rainfall that turns out to be inadequate. These are adaptations over and above iteroparity as such, which is a simpler strategy.

Uncertainty and Parental Care. The effect of uncertainty on degree and distribution of parental investment varies with the type of unpredictability. Some kinds of uncertainty, such as prey availability, apparently can be ameliorated by increased parental investment. Types of uncertainty arising from biotic factors, rather than physical factors, comprise most of this category. Thus, vertebrate predators as a rule should show lengthened juvenile life and high degree of parental care because the biggest and best-taught offspring are at an advantage.

Uncertainties which are catastrophic or otherwise not density dependent appear to favor minimization or delay of parental investment such that the cost of loss at any point before the termination of parental care is minimized. The limited distribution of parental investment in desert amphibians supports this suggestion (fig. 7-3), and it appears to be true not only for anurans, in which parental care varies but is

Table 7-1. *Habitat, Clutch, and Egg Sizes of Various Anurans*

Species	Habitat[a]	Adult Size (mm)	Site of Deposition[c]	Number of Eggs[d]
Ascaphidae				
Ascaphus truei	1D	30–40	2b	28–50/
Leiopelma hochstetteri	8D		4f	6–18/
Pelobatidae				
Scaphiopus bombifrons	1B	35	2a	10–250/ 10–50
S. couchi	1B	80	2a	350–500/ 6–24
S. hammondi	1B	38	2a	300–500/24
Bufonidae				
Bufo alvarius	1B	180	1,2a	7,500– 8,000/
B. boreas	1F	95	2a	16,500/
B. cognatus	1U	Ub	1,2u,2u	30,000/
B. punctatus	1G	55	2a	30–5,000/ 1–few

[a]Habitat: 1 = North America   A = Temporary ponds
             2 = Central America B = Permanent water, xeric areas
             3 = South America   C = Permanent water, mesic areas
             4 = Europe   D = Permanent streams
             5 = Asia   E = Caves
             6 = Africa   F = Mesic   F+ = Cloud or tropical rain forest
             7 = Australia   G = Grasslands, savannahs, or subhumid corridor
             8 = New Zealand
[b]Size of adult female.

Egg Size (mm)	Time to Hatch (hours)	Time to Metamorphose (days)	Time to Mature (years)	Reference
4.0–5.0	720	365+		Noble and Putnam, 1931; Slater, 1934; Stebbins, 1951
	30 days[e]			
	<48	36–40		Stebbins, 1951
.4–1.6	9–72	18–28		Ortenburger and Ortenburger, 1926; Stebbins, 1951; Gates, 1957
.0–1.62	38–120	51		Little and Keller, 1937; Stebbins, 1951; Sloan, 1964
.4		30		Stebbins, 1951; Mayhew, 1968
.5–1.7	48			Stebbins, 1951
.2	53	30–45		Stebbins, 1951
.0–1.3	72	40–60		Stebbins, 1951

[c]Deposition site: 1 = Temporary ponds   3b = Burrows, not requiring rain to hatch
                    2a = Permanent ponds   4a = Terrestrial (seeps, etc.)
                    2b = Permanent streams   4b = On leaves above water
                    3a = Burrows, requiring   4c = On submerged leaves
                         rain to hatch   5 = With parent: brood pouch, on back, etc.

[d]When eggs are laid in several clusters, figures represent total number laid/number per cluster.
[e]Larval development completed in egg.
[f]Tending behavior.   [h]Tadpoles burrow to water.
[g](W): winter (S): summer.   [i]Female digs tunnel to water.

Species	Habitat[a]	Adult Size (mm)	Site of Deposition[c]	Number of Eggs[d]
B. woodhousei	1G	130	1	25,600/
B. compactilis	1G	70	1	
B. microscaphus	1B	65	1	several thousand
B. regularis	6G	65[b]	1	23,000
B. rangeri	6G	105[b]	1,2a	
B. carens	6G	74–92[b]		10,000
B. angusticeps	6	65		650–850
B. gariepensis	6	55		100+
B. vertebralis	6	30		
Ansonia muellari	5D	31[b]		150
Phrynomeridae				
Phrynomerus bifa- sciatus bifasciatus	6G	65	1,2a	400–1,500
Microhylidae				
Gastrophryne carolinensis	1,2	20	2a	850
G. mazatlanensis	3	20	4	175–200
▉▉▉▉ ▉▉▉ ▉▉ spersus adspersus	▉	▉▉	1a,3b	▉▉ ▉▉

[a]Habitat: 1 = North America   A = Temporary ponds
            2 = Central America B = Permanent water, xeric areas
            3 = South America   C = Permanent water, mesic areas
            4 = Europe   D = Permanent streams
            5 = Asia   E = Caves
            6 = Africa   F = Mesic   F+ = Cloud or tropical rain forest
            7 = Australia   G = Grasslands, savannahs, or subhumid corridor
            8 = New Zealand
[b]Size of adult female.

Egg Size (mm)	Time to Hatch (hours)	Time to Metamorphose (days)	Time to Mature (years)	Reference
.0–1.5	48–96	34–60		Mayhew, 1968; Blair, 1972
.4	48			Stebbins, 1951
.75–1.9				Stebbins, 1951
.0	24–48	72–143		Power, 1927; Wager, 1965; Stewart, 1967
.3	96	35–42		Stewart, 1967
.6	72–96			Stewart, 1967
2.0				Wager, 1965
2.2	48			Wager, 1965
<1.0				Wager, 1965
2.15				Inger, 1954
.3–1.5	96	30		Stewart, 1967
	48	20–70	2	Stebbins, 1951
.2–1.4				Stebbins, 1951
.5		28–42 days[e]		Wager, 1965

[c]Deposition site:

1	= Temporary ponds	3b = Burrows, not requiring rain to hatch
2a	= Permanent ponds	4a = Terrestrial (seeps, etc.)
2b	= Permanent streams	4b = On leaves above water
3a	= Burrows, requiring	4c = On submerged leaves
	rain to hatch	5 = With parent: brood pouch, on back, etc.

[d]When eggs are laid in several clusters, figures represent total number laid/number per cluster.
[e]Larval development completed in egg.
[f]Tending behavior.   [h]Tadpoles burrow to water.
[g](W): winter (S): summer.   [i]Female digs tunnel to water.

Species	Habitat[a]	Adult Size (mm)	Site of Deposition[c]	Number of Eggs[d]
B. a. pentheri	6	38	4a,3b	20
Hypopachus variolosus	2G	29–53[b]	1	30–50

Ranidae				
Pyxicephalus adspersus	6G	115	1	3,000–4,000
P. delandii	6G	65	1	2,000–3,000
P. natalensis	6G	51	1	hundreds / 1–6
Ptychadena anchietae	6G	48–58[b]	1	200–300
P. oxyrhynchus	6G	57[b]	1	300–400
P. porosissima	6G	44[b]	1	?/1
Hildebrandtia ornata	6G	63.5	2	?/1
Rana fasciata fuellborni	6C	44.5[b]	4a	64/1–12
R. f. fasciata	6G	51	1,2	?/1
R. angolensis	6	76		thousands
R. fuscigula	6	127	2	1,000– 15,000
R. wageri	6	51[b]	4c	120–1,000/ 12–100

[a]Habitat: 1 = North America   A = Temporary ponds
2 = Central America   B = Permanent water, xeric areas
3 = South America   C = Permanent water, mesic areas
4 = Europe   D = Permanent streams
5 = Asia   E = Caves
6 = Africa   F = Mesic   F+ = Cloud or tropical rain forest
7 = Australia   G = Grasslands, savannahs, or subhumid corridor
8 = New Zealand
[b]Size of adult female.

Egg Size (mm)	Time to Hatch (hours)	Time to Metamorphose (days)	Time to Mature (years)	Reference
5.0	28–42 days[e]			Wager, 1965
	24			Wager, 1965
2.0	48	49		Stewart, 1967
1.5	72	35		Wager, 1965
1.2	96			Wager, 1965; Stewart, 1967
1.0	30			Wager, 1965; Stewart, 1967
1.3	48	42–56		Wager, 1965
1.0	48			Wager, 1965
1.4				Wager, 1965
2.0–3.0		730		Stewart, 1967
1.65		28–35		Wager, 1965
1.5	168			Wager, 1965
1.5	168–240	1,095		Wager, 1965
2.8	192–216			Wager, 1965

[c]Deposition site:
  1 = Temporary ponds
  2a = Permanent ponds
  2b = Permanent streams
  3a = Burrows, requiring rain to hatch
  3b = Burrows, not requiring rain to hatch
  4a = Terrestrial (seeps, etc.)
  4b = On leaves above water
  4c = On submerged leaves
  5 = With parent: brood pouch, on back, etc.

[d]When eggs are laid in several clusters, figures represent total number laid/number per cluster.
[e]Larval development completed in egg.
[f]Tending behavior.
[g](W): winter (S): summer.
[h]Tadpoles burrow to water.
[i]Female digs tunnel to water.

Species	Habitat[a]	Adult Size (mm)	Site of Deposition[c]	Number of Eggs[d]
R. grayi	6B	45	3a	few hundred/1–few
R. catesbiana	1	205	2	10,000–25,000
R. pipiens	1	90	2	1,200–6,500
R. temporaria	1		1	1,500–4,000
R. tarahumarae	1,2	115	1,2	2,200
R. aurora aurora	1C	102	2a	750–1,300
R. a. cascadae	1C	95	2a	425
R. a. dratoni	1C	95	2a	2,000–4,000
R. boylei	1B,C	70	2a, b	900–1,000
R. clamitans	1B,C	102	2a, b	1,000–5,000
R. pretiosa pretiosa	1F	90	2	1,100–1,500
R. p. lutiventris	1F	90	2	2,400
R. sylvatica	1F	60	2a,1	2,000–3,000
Phrynobatrachus natalensis	6G	28–30[b]	1	200–400/25–50
P. ukingensis	6G	16[b]	1	
Anhydrophryne rattrayi	6	20[b]	3b	11–19
Natalobatrachus bonegergi	6F	38	4b	75–100
Arthroleptis wahlbergii	6	29–44[b]	3b	100/33

[a]Habitat: 1 = North America   A = Temporary ponds
            2 = Central America B = Permanent water, xeric areas
            3 = South America   C = Permanent water, mesic areas
            4 = Europe   D = Permanent streams
            5 = Asia   E = Caves
            6 = Africa   F = Mesic   F+ = Cloud or tropical rain forest
            7 = Australia   G = Grasslands, savannahs, or subhumid corridor
            8 = New Zealand
[b]Size of adult female.

Egg Size (mm)	Time to Hatch (hours)	Time to Metamorphose (days)	Time to Mature (years)	Reference
1.5	5–10	90–120		Wager, 1965
1.3	4–5	120–365	2–3	Stebbins, 1951
1.7	312–480	60–90	1–3	Stebbins, 1951
	336–504	90–180	3–5	Stebbins, 1951
2–2.2				Stebbins, 1951
3.04	192–480		3–4	Stebbins, 1951
2.25	192–480			Stebbins, 1951
2.1	192–480			Stebbins, 1951
2.2		90–120		Stebbins, 1951
1.5	72–144	90–360		Stebbins, 1951
2–2.8	96		2+	Stebbins, 1951
1.97				Stebbins, 1951
1.7–1.9	336–504	90		Stebbins, 1951
1.0(W)g 0.7(S)	48	28		Wager, 1965; Stewart, 1967
0.9		35		Stewart, 1967
2.6	28 dayse			Wager, 1965
2.0	144–240	270		Wager, 1965; Stewart, 1967
2.0		e		Stewart, 1967

cDeposition site:
1	= Temporary ponds	3b	= Burrows, not requiring rain to hatch
2a	= Permanent ponds	4a	= Terrestrial (seeps, etc.)
2b	= Permanent streams	4b	= On leaves above water
3a	= Burrows, requiring rain to hatch	4c	= On submerged leaves
		5	= With parent: brood pouch, on back, etc.

dWhen eggs are laid in several clusters, figures represent total number laid/number per cluster.
eLarval development completed in egg.
fTending behavior.
g(W): winter (S): summer.
hTadpoles burrow to water.
iFemale digs tunnel to water.

Species	Habitat[a]	Adult Size (mm)	Site of Deposition[c]	Number of Eggs[d]
A. wageri	6	25	3b	11–30
Arthroleptella lightfooti	6F	20	3b	40/5–8
A. wahlbergi	6F	28	3b	11–30
Cacosternum n. nanum	6G	20	4c	8–25/5–8
Chiromantis xerampelina	6F	60–87[b]	4b	150
Hylambates maculatus	6B	54–70[b]	4c	few hundred/1
Kassina wealii	6	40	1,4c	500/1
K. senegalensis	6	35–43[b]	1	400/1–few
Hemisus marmoratum	6F	38[b]	3fh	200
H. guttatum	6F	64[b]	3fi	2,000
Leptopelis natalensis	6F	64	4a	200
Afrixalus spinifrons	6F	22	4c	?/10–50
A. fornasinii	6F	30–40[b]	4b	40
Hyperolius punticulatus	6	32–43[b]	1	?/19
H. pictus	6	??	1b	?/00–90
H. tuberilinguis	6	36–39[b]	4b	350–400

[a]Habitat:

1 = North America	A = Temporary ponds
2 = Central America	B = Permanent water, xeric areas
3 = South America	C = Permanent water, mesic areas
4 = Europe	D = Permanent streams
5 = Asia	E = Caves
6 = Africa	F = Mesic F+ = Cloud or tropical rain forest
7 = Australia	G = Grasslands, savannahs, or subhumid corridor
8 = New Zealand	

[b]Size of adult female.

Egg Size (mm)	Time to Hatch (hours)	Time to Metamorphose (days)	Time to Mature (years)	Reference
2.5	28 days[e]			Wager, 1965
4.5	10 days[e]			Stewart, 1967
2.5		[e]		Wager, 1965
0.9	48	5		Wager, 1965
1.8	120–144			Wager, 1965
1.5	144	300		Wager, 1965; Stewart, 1967
2.4	144	60		Wager, 1965; Stewart, 1967
1.5	144	90		Stewart, 1967
2.0	240			Wager, 1965
2.5				Wager, 1965
3.0				Wager, 1965
1.2	168	42		Wager, 1965
1.6–2.0				Wager, 1965; Stewart, 1967
2.5				Stewart, 1967
2.0	432	56		Stewart, 1967
1.3–1.5	96–120	60		Stewart, 1967

[c]Deposition site:
1 = Temporary ponds
2a = Permanent ponds
2b = Permanent streams
3a = Burrows, requiring rain to hatch
3b = Burrows, not requiring rain to hatch
4a = Terrestrial (seeps, etc.)
4b = On leaves above water
4c = On submerged leaves
5 = With parent: brood pouch, on back, etc.

[d]When eggs are laid in several clusters, figures represent total number laid/number per cluster.
[e]Larval development completed in egg.
[f]Tending behavior.
[g](W): winter (S): summer.
[h]Tadpoles burrow to water.
[i]Female digs tunnel to water.

Species	Habitat[a]	Adult Size (mm)	Site of Deposition[c]	Number of Eggs[d]
H. pusillus	6	17–21[b]	1,2*a*	500/1–76
H. nasutus nasutus	6	20.6–23.8[b]	2	200/2–20
H. marmoratus nyassae	6	29–31[b]	2*a*	370
H. horstocki	6		2*a*	?/10–30
H. semidiscus	6	35	2*a*, 4*c*	200/30
H. verrucosus	6	29	2*a*	400/4–20
Leptodactylidae				
Eleutherodactylus rugosus	2G		1	several thousand
Limnodynastes tasmaniensis	7F	39.4[b]		1,100
L. dorsalis dumerili	7F	61.5[b]		3,900
Leichriodus fletcheri	7	46.5[b]		300
Adelotus brevus	7	33.5[b]		270
Philoria frosti	7	49.2[b]		95
Helioporus albopunctatus	7	73.3[b]		480
H. eyrei	7F	51.0[b]		605–670
H. psammophilis	7	42–52		160

[a]Habitat: 1 = North America   A = Temporary ponds
         2 = Central America B = Permanent water, xeric areas
         3 = South America   C = Permanent water, mesic areas
         4 = Europe   D = Permanent streams
         5 = Asia   E = Caves
         6 = Africa   F = Mesic   F+ = Cloud or tropical rain forest
         7 = Australia   G = Grasslands, savannahs, or subhumid corridor
         8 = New Zealand
[b]Size of adult female.

Egg Size (mm)	Time to Hatch (hours)	Time to Metamorphose (days)	Time to Mature (years)	Reference
1.4–1.5	120	42		Stewart, 1967
0.8–2.2	120			Wager, 1965; Stewart, 1967
2.0	192			Wager, 1965; Stewart, 1967
1.0				Wager, 1965
1.0	108	60		Wager, 1965
1.3				Wager, 1965
4.0	24			Wager, 1965
1.47				Martin, 1967
1.7				Martin, 1967
1.7				Martin, 1967
1.5				Martin, 1967
3.9				Martin, 1967
2.75				Main, 1965; Lee, 1967
2.50–3.28				Main, 1965; Lee, 1967; Martin, 1967
3.75				Lee, 1967

cDeposition site:
- 1 = Temporary ponds
- 2a = Permanent ponds
- 2b = Permanent streams
- 3a = Burrows, requiring rain to hatch
- 3b = Burrows, not requiring rain to hatch
- 4a = Terrestrial (seeps, etc.)
- 4b = On leaves above water
- 4c = On submerged leaves
- 5 = With parent: brood pouch, on back, etc.

dWhen eggs are laid in several clusters, figures represent total number laid/number per cluster.
eLarval development completed in egg.
fTending behavior.
g(W): winter (S): summer.
hTadpoles burrow to water.
iFemale digs tunnel to water.

Species	Habitat[a]	Adult Size (mm)	Site of Deposition[c]	Number of Eggs[d]
H. barycragus	7	68–80		430
H. inornatus	7	55–65		180
Crinea rosea	7F	24.8[b]		26–32
C. leai	7F	21.1[b]		52–96
C. georgiana	7F	21.1[b]		70
C. insignifera	7F	19–21[b]		

Hylidae

Hyla arenicolor	1	37	1	several hundred/1
H. regilla	1	55	1	500–1,250/ 20–25
H. versicolor	1		2	1,000–2,000
H. verrucigera	2		1	200
H. lancasteri	2F+	41.1[b]	2b,4b	20–23
H. myotympanum	2F+	51.6[b]		120
H. thorectes	2F+	70[b]	2	10
H. ebracata	2F+	36.5[b]	4b	24–76
H. rufelita	2F	60[b]	2	75–80
H. loquax	2F	45[b]	2	250
H. crepitans	3G	59.6[b]	1	
H. pseudopuma	2F	44.2[b]	4b	?/10
H. tica	2F+	38.9		

[a]Habitat:
- 1 = North America
- 2 = Central America
- 3 = South America
- 4 = Europe
- 5 = Asia
- 6 = Africa
- 7 = Australia
- 8 = New Zealand

- A = Temporary ponds
- B = Permanent water, xeric areas
- C = Permanent water, mesic areas
- D = Permanent streams
- E = Caves
- F = Mesic   F+ = Cloud or tropical rain forest
- G = Grasslands, savannahs, or subhumid corridor

[b]Size of adult female.

Egg Size (mm)	Time to Hatch (hours)	Time to Metamorphose (days)	Time to Mature (years)	Reference
2.60				Lee, 1967
3.75				Lee, 1967
2.35		60+ days[e]		Main, 1957
1.66–2.03		149–174 days[e]	2	Main, 1957
0.97–1.3		130+ days[e]	1	Main, 1957
				Main, 1957
2.1		40–70		Stebbins, 1951
1.3	168–336		2	Stebbins, 1951
	96–120	45–65	1–3	Stebbins, 1951
2.0		89		Trueb and Duellman, 1970
5.0				Duellman, 1970
2.25				Duellman, 1970
4.22				Duellman, 1970
1.2–1.4				Duellman, 1970; Villa, 1972
1.8				Villa, 1972
				Villa, 1972
1.8				Villa, 1972
1.71	24	65–69		Villa, 1972
2.0				Villa, 1972

[c]Deposition site: 1 = Temporary ponds   3b = Burrows, not requiring rain to hatch
                      2a = Permanent ponds   4a = Terrestrial (seeps, etc.)
                      2b = Permanent streams 4b = On leaves above water
                      3a = Burrows, requiring 4c = On submerged leaves
                         rain to hatch   5 = With parent: brood pouch, on back, etc.

[d]When eggs are laid in several clusters, figures represent total number laid/number per cluster.

[e]Larval development completed in egg.

[f]Tending behavior.   [h]Tadpoles burrow to water.

[g](W): winter (S): summer.   [i]Female digs tunnel to water.

Species	Habitat[a]	Adult Size (mm)	Site of Deposition[c]	Number of Eggs[d]
Agalychnis colli-dryas	2	71	4b	40–110/ 11–78
A. annae	2	82.9[b]	4b	47–162
A. calcarifer	2	65.0[b]	4b	16
Smilisca cyanosticta	2F+	70[b]	2	1,147
S. baudinii	2G	76–90	1	2,620–3,32●
S. phaeola	2G	80		1,870–2,01●
Pachymedusa dacnicolor	2G	103.6[b]	4b	100–350
Hemiphractus panimensis	2F	58.7[b]	5f	12–14
Gastrotheca ceratophryne	2F	74.2	5f	9
Centrolenellidae				
Centrolenella fleischmanni	2F	19.2	4b	17–28

[a]Habitat: 1 = North America   A = Temporary ponds
2 = Central America   B = Permanent water, xeric areas
3 = South America   C = Permanent water, mesic areas
4 = Europe   D = Permanent streams
5 = Asia   E = Caves
6 = Africa   F = Mesic   F+ = Cloud or tropical rain forest
7 = Australia   G = Grasslands, savannahs, or subhumid corridor
8 = New Zealand
[b]Size of adult female.

generally low, but also for groups with high parental care, such as mammals. For example, marsupials have flourished in uncertain desert environments in central Australia where indigenous and introduced eutherians have not, even though the eutherian species prevail in areas of more predictable climate. In uncertain areas a premium

Egg Size (mm)	Time to Hatch (hours)	Time to Metamorphose (days)	Time to Mature (years)	Reference
2.3–5.0	96–240	50–80		Duellman, 1970; Villa, 1972
3.41				Villa, 1972
3.5				Villa, 1972
1.22				Duellman, 1970
.3				Trueb and Duellman, 1970
				Duellman, 1970
				Duellman, 1970
3.0				Duellman, 1970
2.0				Duellman, 1970
.5	24	9		Villa, 1972

cDeposition site:
- 1 = Temporary ponds
- 2a = Permanent ponds
- 2b = Permanent streams
- 3a = Burrows, requiring rain to hatch
- 3b = Burrows, not requiring rain to hatch
- 4a = Terrestrial (seeps, etc.)
- 4b = On leaves above water
- 4c = On submerged leaves
- 5 = With parent: brood pouch, on back, etc.

dWhen eggs are laid in several clusters, figures represent total number laid/number per cluster.
eLarval development completed in egg.
fTending behavior.
g(W): winter (S): summer.
hTadpoles burrow to water.
iFemale digs tunnel to water.

is set on strategies which will make breeding response facultative and reduce the cost of loss of offspring at any point. Facultative, rather than seasonal, delayed implantation (Sharman, Calaby, and Poole, 1966) and anoestrus condition during drought (Newsome, 1964, 1965, 1966) are examples. Also, I think, is the shape of the parental investment curve for marsupials, which is depressed to a remarkable degree in the initial stages (my unpublished data). This whole constellation of attributes provides facultativeness of response, capabilities for quick initiation of new investments, and less expense of termination at any point. While the classical arguments about marsupial proliferation in Australia have claimed that introduced eutherians "outcompete" marsupials (Frith and Calaby, 1969), they are probably able to do so only because they evolved their reproductive behavior in other kinds of environments. Most Australian environments may have consistently favored marsupialism over any step-by-step transitions toward placentalism. It may be worthwhile to reexamine the question in the light of a new framework.

Distribution

A third important environmental aspect is patchiness or graininess. Wet tropical areas, seemingly ideal from an amphibian's point of view, are basically rather fine grained environments. For instance, ponds, fields, and forest areas may interdigitate so that a single frog spends some time in each and may spend time in more than one pond. From an amphibian's point of view, most deserts are comparatively coarse grained. This does not mean that all the environmental patches are physically large (as may be implied in Levins's [1968] discussion) but that the suitable patches, of whatever size, are likely to be separated by large unsuitable or uninhabitable areas. Thus an individual is likely to spend its entire life in the same patch. For amphibians, widely separated permanent water holes in desert environments are islands and subject to the same selective pressures (MacArthur and Wilson, 1967).

Degrees of patchiness will have two major sorts of effects, on divergence rates and life-history strategies. In a coarse-grained or island model, as in the desert I have described, rates of speciation and extinction will both be higher than in a fine-grained environment. Thus, in some uncertain environments, if they are continually minimally inhabitable and also coarse grained, speciation and extinction rates, contrary to Slobodkin and Sanders's (1969) prediction, may be higher than in predictable environments, if those predictable areas are fine grained. This point, not considered by Slobodkin and Sanders, was raised by Lewontin (1969). Environmental uncertainty will affect populations in the coarse-grained situation much more than those in the fine-grained areas to the extent that there are differences in population sizes and isolation of populations. Slobodkin and Sanders considered only predictability, but predictability and patchiness, and their interaction, will influence the rate of speciation.

In very coarse grained models, because isolation is much more complete than in the fine-grained situation, immigration and emigration may be virtually nonexistent. The number of species in any suitable grain at any time will depend on infrequent past immigrations and will be lower than in the fine-grained model. Selection will be strong on several parameters, to be discussed below, but may be relaxed on characters, such as premating isolating mechanisms. Selection on these characters will be strongest in the fine-grained model where the number of sympatric species is higher. The desert coarse-grained situation is a model for the occurrence of character release (MacArthur and Wilson, 1967; Grant, 1972): populations founded by few individuals and on which selection on interspecific discrimination is relaxed. Thus, in the isolated desert populations described, one might predict that the variations in call characters (in males) and in call discrimination (in females) would be greater.

The distribution of suitable resources and the duration of this distribution will affect strategies of dispersal and competition. While density-dependent effects will operate here, the "r" and "K" parameters of Pianka (1970) and others are not sufficient indicators—a point made by Wilbur et al. (1974) for other groups of organisms.

Consider a pond suitable for breeding: it may be effectively isolated from other suitable areas, or other good ponds may be close or easy to reach. Dispersal ability will evolve to the degree that the cost-benefit ratio is favorable between the relative goodness of another pond and the risk incurred in getting there. Goodness relative to the home pond may be measured by a number of criteria: physical parameters, amount of competition from other species, and other conspecifics (Wilbur et al., 1974), amount of predation, and so on. The cost of reaching another pond and the probability of success in doing so may be correlated with distance, but other classic "barriers" (mountains, very dry areas) are also relevant. Both distance and barriers of low humidity and little free water are likely to be greater in arid regions than in tropical and temperate mesic areas.

If ponds are not totally isolated from each other and are relatively unchanging in "value," migration strategies will be more favored in finer-grained areas because the cost of migration is lower. If ponds are not isolated from each other, and their relative values fluctuate, the evolution of emigration strategies will depend in part on the persistence of ponds relative to the generation length of the frog. If ponds are temporary, and others are likely to be available, migration will be advantageous. The longer ponds last, the closer the situation approaches the "permanent pond" situation, where migration will be favored only in periods of high local population density. Some invertebrate groups, such as migratory locusts and crickets (Alexander, 1968), show phenotypic flexibility supporting this generalization; they increase the proportion of long-winged migratory offspring as the habitat deteriorates and in periods of high population density. Frog morphology does not alter in a comparable way, but dispersal behavior may show flexibility. I know of no pertinent data or studies, however.

In good patches like permanent waters, isolated from others, emigration will be disfavored. Increased parental investment will be favored only when it increases predictability in ways relevant to offspring success. Examination of table 7-1 shows that species with parental care and species laying large-yolked eggs occur in tropical and temperate areas but not generally in unpredictable areas. Since some of these species lay foamy masses not permeable to water, the aridity of desert areas alone is not sufficient to explain this distribution of strategies.

Two arid region anopia in the parental investment in the form of larger or protected eggs. As previously described, Eleutherodactylus latrans females lay about fifty large eggs of 6–7.5 mm diameter on land or in caves (table 7-1; Stebbins, 1951); the males may guard the eggs. Since this frog lives largely in caves and rocky crevices, the microenvironment is far more stable and predictable than the zoogeography would suggest. The Australian Pseudophryne occidentalis lives by permanent waters with muddy rather than sandy soils. Eggs are laid in mud burrows near the edge of the water (Main, 1965). In both cases it appears that the nature of mortality is such that increased parental investment is successful. This may be related to the relatively higher physical stability of the microhabitat when compared to desert environments in general. The proportion of mortality due to catastrophes which parental care is ineffective to combat is relatively lower.

Mortality

Mortality may arise from a number of factors: foot shortages, predators (including parasites and diseases), and climate. An important consideration in what life-history strategy will prevail is whether the mortality is random (unpredictable) or nonrandom (predictable). Any cause of mortality could be either random or nonrandom in its effects, but mortality from biotic causes is probably less often random than mortality from physical factors and may be more effectively countered by strategies of parental investment.

Catastrophic mortality, which is essentially random rather than selective (even though it may be density dependent), will be more frequent in the coarse-grained desert environments I have described than in the tropics. An example would be heavy sudden floods which frequently occur after heavy rains in areas like central Australia and the southwestern United States. This kind of flood may wash eggs, tadpoles, and adults to flood-out areas which then dry up. The result may be devastating sporadic mortality for populations living in the path of such floods. Further, in terms of the animals themselves, environments may be predictable for certain stages in the life history and unpredictable for others. In animals like amphibians with complex metamorphosis, this difference can be particularly significant.

If any stage encounters significant uncertainty, one of two strategies should evolve: physical avoidance, such as hiding or development of protection in that stage, or a shift in life history to spend minimal time in the vulnerable stage (table 7-2). If survivorship is high for adults but uncertain and sometimes very low for tadpoles, one predicts strategies of: (a) long adult life, iteroparity, and reduced investment per clutch; (b) long egg periods and short tadpole periods; or (c) increased parental investment through hiding or tending behavior. Evolution of behavior like that of *Rinoderma darwini* may result from such pressure. The males appear to guard the eggs; when development reaches early tadpole stage, the males snap up the larvae, carrying them in the vocal sac until metamorphosis. Perhaps the extreme case is represented by the African *Nectophrynoides*, in which birth is viviparous.

In temporary waters in desert environments much uncertainty will be concentrated on aquatic stages, and two principal strategies should be evident in desert amphibians: increased iteroparity, longer adult life, and lower reproductive effort per clutch; and shifts in time spent in different stages, reducing time spent in the vulnerable stages. Short, variable lengths in egg and juvenile stages (table 7-1) will result.

Even in climatically more predictable areas, uncertainty of mortality may be concentrated on one stage. In some temperate urodele forms, Salthe (1969) suggested that success at metamorphosis correlated with size—that larger offspring were more successful. This in turn selected for lengthened time spent in aquatic stages.

Some generalizations are apparent from table 7-2. The important differences appear to be between uncertainty in juvenile stages and adult stages. All conditions of uncertain adult survival will lead to concentration of reproductive effort in one or a few clutches (semelparity or reduction of iteroparity). Uncertainty of survivorship in adult stages when combined with high predictability in juvenile stages may lead to the extreme conditions of neoteny and paedogenesis. Uncertainty in either or both juvenile stages leads to increased iteroparity and reduced reproductive effort per clutch.

Predation

Because predation is usually nonrandom, its effects on prey life histories will frequently differ from the effects of climate and other sources of mortality. An important point frequently overlooked is that, because predation and competition arise from biotic components of the system, they are not simply subsets of uncertainty. Their effects are more thoroughly related to density-dependent parameters. Some strategies will be effective which would not be advantageous in situations rendered uncertain solely by physical factors. Consider predation: strategies frequently effective in reducing predation-caused un-

Table 7-2. *Relative Uncertainty in Different Life-History Stages and Strategies of Selective Advantage*

Likelihood of Survival			Strategy
Egg	Tadpole	Adult	
high	high	low	semelparity or reduced iteroparity; large numbers of small eggs; no parental care
low	high	low	semelparity or reduced iteroparity; neoteny; quick hatching
high	low	low	semelparity or reduced iteroparity; large numbers of small eggs; no parental care; quick metamorphosis
high	low	high	iteroparity; large eggs, fewer eggs; avoidance of aquatic tadpole stage; parental care of tadpoles
low	high	high	iteroparity; tending, hiding of eggs; fewer eggs; viviparity
low	low	high	iteroparity; parental care, tending strategies; viviparity

certainty are those of spatial (Hamilton, 1971) and temporal clumping, increased parental investment (Trivers, 1972), and allelochemical effects. These strategies would be far less effective in increasing predictability of an environment rendered uncertain by physical factors.

Predation pressure may lead to hiding or tending eggs and consequent lowering of clutch size. Whether this is true or whether responses of increased fecundity (Porter, 1972; Szarski, 1972) prevail will depend on the nature of the predation. In the unusual case of a predator whose effect is limited, such as one which could eat no more than x eggs per nest, parents would gain by increased fecundity, making $(x+2)$ rather than $(x+1)$ eggs. However, m, the genotypic rate of increase, will be higher for these more fecund genotypes even in the absence of predation. Further, an increase in numbers of eggs laid implies either smaller eggs (in which case the predator may be able to eat $[x+2]$ eggs) or an increase in the size of the parent. In most cases, high fecundity carries a greater risk under increased predation—for example, by laying more eggs which are then lost or, in species like altricial birds with parental care, by incurring greater risk attempting to feed more offspring if they are not protected. In these cases, lowered fecundity and increased parental investment in caring for fewer eggs will be favored.

The strategies of hiding or protection and life-history shifts, which may follow from increased uncertainty in any stage, are also favored in the special case of uncertainty induced by predation. Predation concentrated on certain stages in the life cycle—on eggs, tadpoles, newly metamorphosed animals, or breeding adults—may lead to (a) quick hatching, tending, or hiding of eggs, as in *Scaphiopus* or *Helioporus* (table 7-1); (b) quick metamorphosis or tending of tadpoles, as in *Rhinoderma*; (c) cryptic behavior by newly metamorphosed animals (many species) or lengthened egg or tadpole stages with consequent greater size (and possibly reduced predation vulnerability) on metamorphosis, as in *Rana catesbiana* (table 7-1); or (d) cryptic behavior by adults or very clumped patterns of breeding behavior.

Length of the breeding season may also be strongly affected by the presence of predation. In fact, I think that the general shape of breeding-curve activities of many vertebrates may be related to predation. Fisher (1000) has shown that, if there is an optimal breeding time, a symmetrical curve will result. While restriction of resource availability, such as food or breeding resources, limits the seasonality of breeding and produces some clumping, such seasonal differences seem not to be sharp enough to explain the extreme temporal clumping of breeding and birth in many species. Temporary ponds of very short duration in arid regions are commonly assumed to show clumping for climatic reasons, but this is not certain; at any rate, the addition of predation to such a system should follow the same pattern as in any seasonal situation. In seasonal conditions a breeding-activity or birth curve may approach a normal curve, perhaps with a slight right-hand skew because earlier birth will give a size and food advantage to offspring and a risk advantage to parents. When predation on breeding adults or new young exists, however, two other pressures may cause both an increased right-hand skew and a sharper peak:

1. The advantage to those individuals which have offspring early before a generalized predator develops a specific search image.

2. The advantage to those individuals which breed and give birth or lay eggs when everyone else does—when, in other words, the predator food market is flooded. This constitutes a temporal "selfish herd" effect (Hamilton, 1971). Thus, if seasonality of resource availability exists so that thoroughly cryptic breeding is not of advantage, the curve of breeding or birth activity will tend under predation pressure to shift from a fairly normal distribution to a kurtotic curve with an abrupt beginning shoulder and a gentler trailing edge.

Despite their importance, predation effects on life histories have largely been ignored. This may be, in part, because the physical factors are so extreme that it seems sufficient to examine their effects on amphibian physiology and survival. Another reason predation effects may be slighted is that one ordinarily sees the end product of organisms which evolved with predation pressure, and the present-day descendents represent the most successful of the antipredation strategies. As a simple example, consider the large variety of substances found in the skin of most amphibians (Michl and Kaiser, 1963). A great variety exists, including such disparate compounds as urea, the bufadienolides, indoles, histamine derivatives, and polypeptides like caerulein (Michl and Kaiser, 1963; Erspamer, Vitali, and Roseghini, 1964; Anastasi, Erspamer, and Endean, 1968; Cei, Erspamer, and Roseghini, 1972; Low, 1972). The production of some of these compounds is energetically expensive; others are costly in terms of water economy (Cragg, Balinsky, and Baldwin, 1961; Balinsky, Cragg, and Baldwin, 1961). Why, then, do so many amphibians produce a wide variety of such costly compounds? Despite wide chemical variety most of these compounds share one striking attribute: they are either distasteful or have unpleasant physiological effects. Most irritate the mucous membranes. Bufadienolides and other cardiac glycosides have digitalislike effects on such predators as snakes as well as on mammals (Licht and Low, 1968). Caerulein differs in only two amino acids from gastrin and has similar effects (Anastasi et al., 1968), including the induction of vomiting.

Although I know of no good study of predation mortality in any desert amphibian, and demography data on amphibians are generally sparse (Turner, 1962), predation has been reported in every life-history stage (Surface, 1913; Barbour, 1934; Brockelman, 1969; Littlejohn, 1971; Szarski, 1972). It is obvious that there is selective advantage to tasting vile or being poisonous, and scattered studies show that successful predators on amphibians show adaptations of increased tolerance (Licht and Low, 1968) or avoidance of the poisonous parts (Miller, 1909; Wright, 1966; Schaaf and Garton, 1970).

Predation concentrated on adults will lead to the success of individuals which show cryptic behavior and color patterns as well as those which concentrate unpleasant compounds in their skins. Particularly poisonous or distasteful individuals with bright or striking color patterns may also be favored (Fisher, 1958). Two apparently opposite breeding strategies may succeed, depending on other factors discussed below. These are cryptic breeding behavior and temporally and spatially clumped breeding behavior.

Several strategies may evolve as a response to predation on eggs: eggs with foam coating, as in a number of *Limnodynastes* species (Martin, 1967, Littlejohn, 1971); eggs containing poisonous substances, as in *Bufo* (Licht, 1967, 1968); eggs hatching quickly, as in *Scaphiopus* (Stebbins, 1951; Bragg, 1965, summarizing earlier papers); and a clumping of egg laying or hiding or tending of eggs, as is done by a number of New World tropical species (table 7-1). If adults become poisonous and effectively invulnerable, they concomitantly become good protectors of the eggs.

The strategies of hiding or tending eggs involve a greater parental investment per offspring and result in a decrease in the total number of eggs laid (figs. 7-1 and 7-2). That a general correlation exists between strategies of parental care and numbers of eggs has been recognized for some time; but no pattern has been recognized, and explanations by herpetologists have verged on the teleological, such as those of Porter (1972).

Figures 7-1, 7-2, and 7-3 show the relationships of egg sige, female size, litter size, and predictability of habitat. Indeed, as the size of egg relative to the female increases, the clutch size decreases (table 7-1, fig. 7-1). This is as expected and correlates with results from other groups (Williams, 1966a, 1966b; Salthe, 1969; Tinkle, 1969). When habitat or egg-laying locality is shown on a graph plotting the ratio of egg size to female size against litter size (fig. 7-3), it is apparent that most of those species showing some increase in parental care, such as laying eggs in burrows or leaves or tending the eggs or tadpoles, lay fewer, larger eggs; these species without exception live in habitats of relatively high environmental predictability—tropical rain forests, caves, and so on (table 7-1). No species laying eggs in temporary ponds show such behavior. The species in areas of high predictability possess a variety of strategies of high parental investment per off-spring. As mentioned above, *Rhinoderma darwini* males carry the eggs in the vocal sac (Porter, 1972, and others). *Leiopelma hochstetteri* eggs are laid terrestrially and tended by one of the parents.

Females of several species of *Helioporus* lay eggs in a burrow excavated by the male, and the eggs await flooding to hatch (Main, 1965; Martin, 1967). Eggs of *Pipa pipa* are essentially tended by the female, on whose back they develop. Barbour (1934) and Porter (1972) reviewed a number of cases of parental tending and hiding strategies.

In situations (such as physical uncertainty or unpredictable predation) where increased parental investment per offspring is ineffective in decreasing the mortality of an individual's offspring, the minimum investment per offspring will be favored. In these cases, individuals which win are those which lay eggs in the peak laying period and in the middle of a good area being used by others. Any approaching predator should encounter someone else's eggs first. This strategy should be common in deserts and indeed appears to be (table 7-1). The costs of playing this temporal and spatial variety of "selfish herd" game (Hamilton, 1971) are that some aspects of intraspecific competition are maximized and predators may evolve to exploit the conspicuous "herd."

Three strategies would appear to be of selective advantage if predation is concentrated on the tadpole stage. One is the laying of larger or larger-yolked eggs producing larger and less-vulnerable tad-

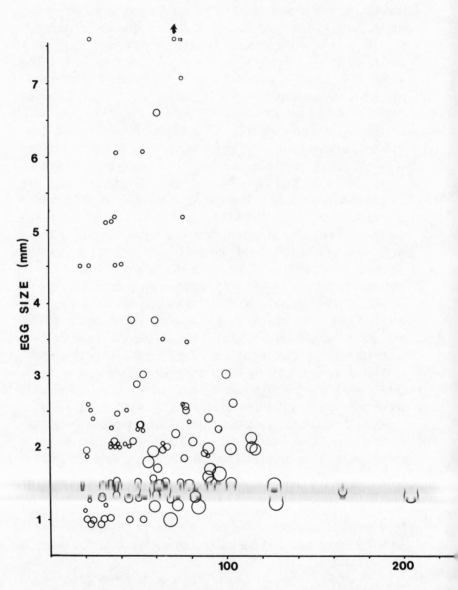

Fig. 7-1. Relationship of egg size to size of adult female for species from table 7-1. Size of circle indicates size of clutch:

o = ⟨ 500   ◯ = 1,000–10,000   ◯ = ⟩ 10,000

◯ = 500–1,000

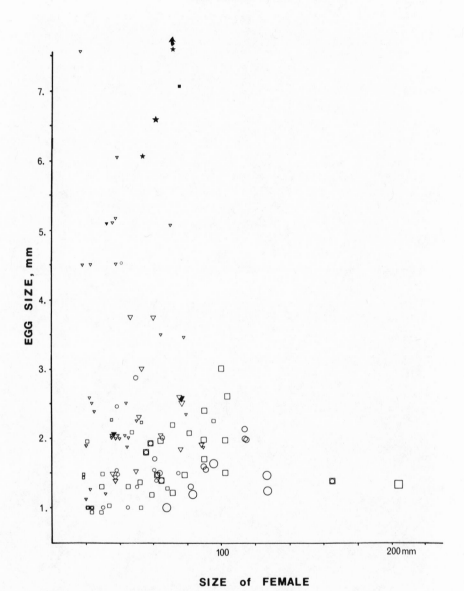

SIZE of FEMALE

Fig. 7-2. *Relationship of egg size to size of adult female. As in figure 7-1, clutch size is shown by size of symbol. Solid symbols indicate tending behavior by a parent. Habitat of eggs:*

    ○ = *temporary water*   Δ = *laid in burrows, wrapped in leaves, etc.*
    □ = *permanent water*   ★ = *carried in brood pouch, on back, in*
                                            vocal sac

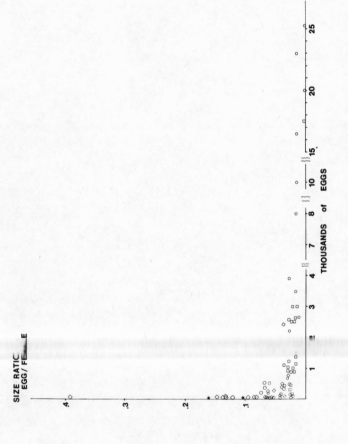

Fig. 7-3. Relationship of egg habitat to clutch size and relative size of eggs.
Habitat of eggs:

○ = temporary water   ◇ = laid on leaves, in burrows
□ = permanent water   ★ = carried by parent

poles at hatching. If seasonal or environmental conditions permit, producing offspring which spend a longer time as eggs may be successful. This would frequently involve strategies of hiding or tending eggs. In some species the entire development is completed while secreted so that on emergence offspring, in fact, are adults (*Leiopelma, Rhinoderma*). A third strategy is that of facultatively quick metamorphosis (*Bufo, Scaphiopus*), a strategy one might also expect to be favored in temporary ponds in desert situations. However, this advantage is balanced by intraspecific competition with its contingent selective advantage on size. So, in fact, what one would predict whenever genetic cost is not too great (Williams, 1966a) are facultative lengths of egg and larval periods and facultative hatching and metamorphosis. Thus, under strong predation and in more uncertain environments, one predicts an increase in facultativeness in these parameters. While this is predictable for both factors, studies of predation have not been able to separate out effects (De Benedictis, 1970). In amphibians of desert temporary ponds, either length of egg and larval periods are short or there is a large variation in reported lengths. Lengths of time to hatch in *S. couchi*, for example, range from 9 hours (Ortenburger and Ortenburger, 1926) to 48–72 hours (Gates, 1957); and in *S. hammondi*, from 38 (Little and Keller, 1937) to 120 hours (Sloan, 1964). The sizes at which these species metamorphose are highly variable (Bragg, 1965), suggesting that the strongest pressures of uncertainty center on the tadpole stages.

If predation is concentrated on juveniles, there will be an advantage to cryptic behavior by newly metamorphosed individuals. If predation is nonrandom and size related (as appears likely at metamorphosis when major predators are fish and other frogs), larger-sized individuals will be favored. If laying larger-yolked eggs results in larger offspring and increased offspring survivorship, this strategy will win. Certainly spending a longer time in the egg and tadpole stages, if predation is not heavily concentrated on these stages, will be favored. In species like *Rana catesbiana*, length of larval life is facultative and, for late eggs, is greater than a year. This appears to be involved with time required to reach a large enough size to be relatively invulnerable as a juvenile. While lengthened juvenile life cannot be selected for directly, conditions like predation on newly metamorphosed individuals are precisely those rendering lengthened periods in the tadpole stage advantageous.

Suggestions for Future Research

We can see that the interplay of these conditions is complex, and it is not necessarily a simple undertaking to predict strategies favored in each situation. Presently observed situations reflect the summation of a number of possibly conflicting selective advantages. Further, even though some biotic factors may be partially predictable from physical factors (e.g., in seasonal situations it is predictable that not only will food and breeding suitability be greater at some periods than others, but also at those same times predation will increase), others are not, and there is no single simple pattern.

The questions raised here are difficult to answer without further data, which are skimpy for anurans. Studies like Tinkle's (1969) on lizards or Inger and Greenberg's (1963) would afford comparative data for examination in the theoretical approach put forward here. For the most part, work on life histories in anurans has been zoogeographic and anecdotal. We haven't asked the right questions. Needed now are comparative studies similar to Tinkle's, between similar species in different habitats, and, in wide-ranging species, conspecific comparisons between habitats. We need to have:

1. Demographic data including length of life, time to maturity, age-specific fecundity, and degree of iteroparity, including number of eggs per clutch and number of clutches per year.
2. Ratio of egg size to female size.
3. Behavior: territoriality, tending behavior. (For example, Porter [1972] reported that *Rhinoderma darwini* males tend eggs that may not be their own. Such genetic altruism seems unlikely and needs further examination.)
4. Within wide-ranging species, comparative studies including, in addition to the above, work on mating-call parameters of males and discrimination of females.

Only when we begin to ask the above kinds of questions will we be able to develop an overall theoretical framework within which to view amphibian life histories. Many of the predictions and speculations discussed here seem obvious or trivial, but perhaps such attempts are necessary first steps toward a conceptual treatment of amphibian life histories.

Summary

Despite their normal requirement for an aqueous environment during the larval stage, a considerable number of amphibian species have adapted successfully to the desert environment. Possible methods of adaptation are considered, and their occurrence is reviewed. A number depend on modification of life histories, and attention is concentrated on these. Success depends on balancing the risk of mortality against the cost of reproductive effort.

Since desert environments are often less predictable than others, life-history strategies must take this uncertainty into account. This implies repeated small but prompt reproductive efforts and long adult life; behavior which enhances positive effects of random events, or reduces their negative effects, will be favored. Reduction in parental investment is generally advantageous in conditions of uncertainty in the physical environment.

From the amphibian point of view, the desert environment is patchy —coarse-grained—with high rates of speciation and extinction. Migration is favored where ponds are temporary and disfavored where they are permanent.

Mortality in the deserts is much more random than in mesic environments where it is dominated by predation. Reduction in the duration of vulnerable stages will then be advantageous. Responses to predation, however, have helped to shape amphibian life histories in the desert, as well as leading to production of noxious substances in many species. Differences in egg size and number per clutch may depend on likelihood of predation as against other hazards.

The importance of increased information about amphibian demography, and aspects of behavior related to it, is emphasized.

References

Alexander, R. D. 1968. Life cycle origins, speciations, and related phenomena in crickets. *Q. Rev. Biol.* 43:1–41.
Anastasi, A.; Erspamer, V.; and Endean, R. 1968. Isolation and

amino acid sequence of caerulein, the active decapeptide of the skin of *Hyla caerulea. Archs Biochem. Biophys.* 125:57–68.

Balinsky, J. B.; Cragg, M. M.; and Baldwin, E. 1961. The adaptation of amphibian waste nitrogen excretion to dehydration. *Comp. Biochem. Physiol.* 3:236–244.

Barbour, T. 1934. *Reptiles and amphibians: Their habits and adaptations*. Boston and New York: Houghton Mifflin.

Bentley, P. J. 1966. Adaptations of Amphibia to desert environments. *Science, N.Y.* 152:619–623.

Bentley, P. J.; Lee, A. K.; and Main, A. R. 1958. Comparison of dehydration and hydration in two genera of frogs (*Helioporus* and *Neobatrachus*) that live in areas of varying aridity. *J. exp. Biol.* 35: 677–684.

Blair, W. F., ed. 1972. *Evolution in the genus "Bufo."* Austin: Univ. of Texas Press.

Bragg, A. N. 1965. *Gnomes of the night: The spadefoot toads*. Philadelphia: Univ. of Pa. Press.

Brattstrom, B. H. 1962. Thermal control of aggregation behaviour in tadpoles. *Herpetologica* 18:38–46.

———. 1963. A preliminary review of the thermal requirements of amphibians. *Ecology* 24:238–255.

Brockelman, W. Y. 1969. An analysis of density effects and predation in *Bufo americanus* tadpoles. *Ecology* 50:632–644.

Cei, J. M.; Erspamer, V.; and Roseghini, M. 1972. Biogenic amines. In *Evolution in the genus "Bufo,"* ed. W. F. Blair. Austin: Univ. of Texas Press.

Cragg, M. M.; Balinsky, J. B.; and Baldwin, E. 1961. A comparative study of the nitrogen excretion in some Amphibia and Reptilia. *Comp. Biochem. Physiol.* 3:227–236.

De Benedictis, P. A. 1970. "Interspecific competition between tadpoles of *Rana pipiens* and *Rana sylvatica*: An experimental field study." Ph.D. dissertation, University of Michigan.

Dole, J. W. 1967. The role of substrate moisture and dew in the water economy of leopard frogs, *Rana pipiens. Copeia* 1967:141–150.

Duellman, W. E. 1970. The hylid frogs of Middle America. *Monogr. Univ. Kans. Mus. nat. Hist.* 1:1–753.

Erspamer, V.; Vitali, T.; and Roseghini, M. 1964. The identification of new histamine derivatives in the skin of *Leptodactylus*. *Archs Biochem. Biophys.* 105:620–629.

Fisher, R. A. 1958. *The genetical theory of natural selection*. 2d rev. ed. New York: Dover.

Frith, H. J., and Calaby, J. H. 1969. *Kangaroos*. Melbourne: F. W. Cheshire.

Gates, G. O. 1957. A study of the herpetofauna in the vicinity of Wickenburg, Maricopa County, Arizona. *Trans. Kans. Acad. Sci.* 60:403–418.

Grant, P. R. 1972. Convergent and divergent character displacement. *J. Linn. Soc. (Biol.)* 4:39–68.

Hamilton, W. D. 1971. Geometry for the selfish herd. *J. theoret. Biol.* 31:295–311.

Heatwole, H.; Blasina de Austin, S.; and Herrero, R. 1968. Heat tolerances of tadpoles of two species of tropical anurans. *Comp. Biochem. Physiol.* 27:807–815.

Hussell, D. J. T. 1972. Factors affecting clutch-size in Arctic passerines. *Ecol. Monogr.* 42:317–364.

Inger, R. F. 1954. Systematics and zoogeography of Philippine Amphibia. *Fieldiana, Zool.* 33:185–531.

Inger, R. F., and Greenberg, B. 1956. Morphology and seasonal development of sex characters in two sympatric African toads. *J. Morph.* 99:549–574.

———. 1963. The annual reproductive pattern of the frog *Rana erythraea* in Sarawak. *Physiol. Zoöl.* 36:21–33.

Klomp, H. 1970. The determination of clutch-size in birds. *Ardea* 58: 1–124.

Lack, D. 1947. The significance of clutch-size. Pts. I and II. *Ibis* 89: 302–352.

———. 1948. The significance of clutch-size. Pt. III. *Ibis* 90:24–45.

Lee, A. K. 1967. Studies in Australian Amphibia. II. Taxonomy, ecology, and evolution of the genus *Helioporus* Gray (Anura: Leptodactylidae). *Aust. J. Zool.* 15:367–439.

Lee, A. K., and Mercer, E. H. 1967. Cocoon surrounding desert-dwelling frogs. *Science, N.Y.* 157:87–88.

Levins, R. 1968. *Evolution in changing environments*. Monographs in Population Biology, 2. Princeton: Princeton Univ. Press.

Lewontin, R. C. 1969. Comments on Slobodkin and Sanders "Contribution of environmental predictability to species diversity." *Brookhaven Symp. Biol.* 22:93.

Licht, L. E. 1967. Death following possible ingestion of toad eggs. *Toxicon* 5:141–142.

———. 1968. Unpalatability and toxicity of toad eggs. *Herpetologica* 24:93–98.

Licht, L. E., and Low, B. S. 1968. Cardiac response of snakes after ingestion of toad parotoid venom. *Copeia* 1968:547–551.

Little, E. L., and Keller, J. G. 1937. Amphibians and reptiles of the Jornada Experimental Range, New Mexico. *Copeia* 1937:216–222.

Littlejohn, M. J. 1967. Patterns of zoogeography and speciation by southeastern Australian Amphibia. In *Australian inland waters and their fauna*, ed. A. H. Weatherley, pp. 150–174. Canberra: Aust. Nat. Univ. Press.

———. 1971. Amphibians of Victoria. *Victorian Year Book* 85:1–11.

Low, B.S. 1972. Evidence from parotoid gland secretions. In *Evolution in the genus "Bufo,"* ed. W. F. Blair. Austin: Univ. of Texas Press.

MacArthur, R. H., and Wilson, E. O. 1967. *The theory of island biogeography*. Monographs in Population Biology, 1. Princeton: Princeton Univ. Press.

Main, A. R. 1957. Studies in Australian Amphibia. I. The genus *Crinia tschudi* in south-western Australia and some species from southeastern Australia. *Aust. J. Zool.* 5:30–55.

———. 1962. Comparisons of breeding biology and isolating mechanisms in Western Australian frogs. In *The evolution of living organisms*, ed. G. W. Leeper. Melbourne: Melbourne Univ. Press.

———. 1965. *Frogs of southern Western Australia*. Perth; West Australian Nat. Club.

———. 1968. Ecology, systematics, and evolution of Australian frogs. *Adv. ecol. Res.* 5:37–87.

Main, A. R., and Bentley, P. J. 1964. Water relations of Australian burrowing frogs and tree frogs. *Ecology* 45:379–382.

Main, A. R.; Lee, A. K.; and Littlejohn, M. J. 1958. Evolution in three genera of Australian frogs. *Evolution* 12:224–233.

Main, A. R.; Littlejohn, M. J.; and Lee, A. K. 1959. Ecology of Australian frogs. In *Biogeography and ecology in Australia*, ed. A. Keast, R. L. Crocker, and C. S. Christian. The Hague: Dr. W. Junk.

Martin, A. A. 1967. Australian anuran life histories: Some evolutionary and ecological aspects. In *Australian inland waters and their fauna*, ed. A. H. Weatherley, pp. 175–191. Canberra: Aust. Nat. Univ. Press.

Mayhew, W. W. 1968. Biology of desert amphibians and reptiles. In *Desert biology*, ed. G. W. Brown, vol. 1, pp. 195–356. New York and London: Academic Press.

Michl, H., and Kaiser, E. 1963. Chemie and Biochemie de Amphibiengifte. *Toxicon* 1963:175–228.

Miller, N. 1909. The American toad. *Am. Nat.* 43:641–688.

Murphy, G. I. 1968. Pattern in life history and the environment. *Am. Nat.* 102:391–404.

Newsome, A. E. 1964. Anoestrus in the red kangaroo, *Megaleia rufa*. *Aust. J. Zool.* 12:9–17.

———. 1965. The influence of food on breeding in the red kangaroo in central Australia. *CSIRO Wildl. Res.* 11:187–196.

———. 1966. Reproduction in natural populations of the red kangaroo *Megaleia rufa* in central Australia. *Aust. J. Zool.* 13:735–759.

Noble, C. K., and Putnam, P. G. 1931. Observations on the life history of *Ascaphus truei* Stejneger. *Copeia* 1931:97–101.

Ortenburger, A. I., and Ortenburger, R. D. 1926. Field observations on some amphibians and reptiles of Pima County, Ariz. *Proc. Okla. Acad. Sci.* 6:101–121.

Pianka, E. R. 1970. On r and K selection. *Am. Nat.* 104:592–597.

Porter, K. R. 1972. *Herpetology*. Philadelphia: W. B. Saunders Co.

Power, J. A. 1927. Notes on the habits and life histories of South African Anura with descriptions of the tadpoles. *Trans. R. Soc. S. Afr.* 14:237–247.

Ruibal, R. 1962a. The adaptive value of bladder water in the toad, *Bufo cognatus*. *Physiol. Zoöl.* 35:218–223.

———. 1962b. Osmoregulation in amphibians from heterosaline habitats. *Physiol. Zoöl.* 35:133–147.

Salthe, S. N. 1969. Reproductive modes and the number and size of ova in the urodeles. *Am. Midl. Nat.* 81:467–490.

Schaaf, R. T., and Garton, J. S. 1970. Racoon predation on the American toad, *Bufo americanus*. *Herpetologica* 26:334–335.

Schmidt, K. P., and Inger, R. F. 1959. Amphibia. *Explor. Parc natn. Upemba Miss. G. F. de Witt* 56.

Sharman, G. B.; Calaby, J. H.; and Poole, W. E. 1966. Patterns of reproduction in female diprotodont marsupials. *Symp. zool. Soc. Lond.* 15:205–232.

Slater, J. R. 1934. Notes on northwestern amphibians. *Copeia* 1934: 140–141.

Sloan, A. J. 1964. Amphibians of San Diego County. *Occ. Pap. S Diego Soc. nat. Hist.* 13:1–42.

Slobodkin, L. D., and Sanders, H. L. 1969. On the contribution of environmental predictability to species diversity. *Brookhaven Symp. Biol.* 22:82–96.

Stebbins, R. C. 1951. *Amphibians of western North America*. Berkeley and Los Angeles: Univ. of Calif. Press.

Stewart, M. M. 1967. *Amphibians of Malawi*. Albany: State Univ. of N.Y. Press.

Surface, H. A. 1913. The Amphibia of Pennsylvania. *Bi-m. zool. Bull. Pa Dep. Agric.* May–July 1913:67–151.

Szarski, H. 1972. Integument and soft parts. In *Evolution in the genus "Bufo,"* ed. W. F. Blair. Austin: Univ. of Texas Press.

Tinkle, D. W. 1969. The concept of reproductive effort and its relation to the evolution of life histories of lizards. *Am. Nat.* 103:501–514.

Trivers, R. L. 1972. Parental investment and sexual selection. In *Sexual selection and the descent of man*, ed. B. Campbell, pp. 136–179. Chicago: Aldine.

Trueb, L., and Duellman, W. E. 1970. The systematic status and life history of *Hyla verrucigera* Werner. *Copeia* 1970:601–610.

Turner, F. B. 1962. The demography of frogs and toads. *Q. Rev. Biol.* 37:303–314.

Villa, J. 1972. *Anfibios de Nicaragua*. Managua: Instituto Geográfico Nacional, Banco Central de Nicaragua.

Volpe, E. P. 1953. Embryonic temperature adaptations and relationships in toads. *Physiol. Zoöl.* 26:344–354.

Wager, V. A. 1965. *The frogs of South Africa*. Capetown: Purnell & Sons.

Warburg, M. R. 1965. Studies on the water economy of some Australian frogs. *Aust. J. Zool.* 13:317–330.

Wilbur, H. M., and Collins, J. P. 1973. Ecological aspects of amphibian metamorphosis. *Science, N.Y.* 182:1305.

Wilbur, H. M.; Tinkle, D. W.; and Collins, J. P. 1974. Environmental certainty, trophic level, and successional position in life history evolution. *Am. Nat.* 108:805–818.

Williams, G. C. 1966a. *Adaptation and natural selection: A critique of some current evolutionary thought.* Princeton: Princeton Univ. Press.

———. 1966b. Natural selection, the costs of reproduction, and a refinement of Lack's principle. *Am. Nat.* 100:687–692.

Wright, J. W. 1966. Predation on the Colorado River toad, *Bufo alvarius. Herpetologica* 22:127–128.

8. Adaptation of Anurans to Equivalent Desert Scrub of North and South America

W. Frank Blair

Introduction

The occurrence of desertic environments at approximately the same latitudes in western North America and in South America provides an excellent opportunity to investigate comparatively the structure and function of ecosystems that have evolved under relatively similar environments. A multidisciplinary investigation of these ecosystems to determine just how similar they are in structure and function is presently in progress under the Origin and Structure of Ecosystems Program of the U.S. participation in the International Biological Program. The specific systems under study are the Argentine desert scrub, or Monte, as defined by Morello (1958) and the Sonoran desert of southwestern North America.

In this paper I will discuss the origins and nature of one component of the vertebrate fauna of these two xeric areas, the anuran amphibians. Pertinent questions are (a) How do the two areas compare in the degree of desert adaptedness of the fauna? (b) How do the two areas compare with respect to the size of the desert fauna? (c) What are the geographical origins of the various components of the fauna? and (d) What are the mechanisms of desert adaptation?

The comparison of the two desert faunas must take into account a number of major factors that have influenced their evolution. The most important among these would seem to be:

1. The nature of the physical environment of physiography and climate
2. The degree of similarity of the vegetation in general ecological aspect and in plant species composition

3. The size of each desert area
4. Possible sources of desert-invading species and the nature of adjacent biogeographic areas
5. The past history of the area through Tertiary and Pleistocene times
6. The evolutionary-genetic capabilities of available stocks for desert colonization

The Physical Environment

As defined by Morello (1958), the Monte extends through approximately 20° of latitude from 24°35′S in the state of Salta to 44°20′S in the state of Chubut and through approximately 7° of longitude from 69°50′W in Neuquen to 62°54′W on the Atlantic coast. The Sonoran desert occupies an area lying approximately between lat. 27° and 34°N and between long. 110° and 116°W (Shelford, 1963, fig. 15-1). Both areas are characterized by lowlands and mountains. The present discussion will deal principally with the lowland fauna.

Rainfall in both of the areas is usually less than 200 mm annually (Morello, 1958; Barbour and Díaz, 1972). Thus, availability of water is the most important factor determining the nature of the vegetation and the most important control limiting the invasion of these areas by terrestrial vertebrates.

The Vegetation

A more precise discussion of the vegetation of the Monte will be found elsewhere in this volume (Solbrig, 1975), so I will point out only that the general aspect is very similar in the two areas. The genera *Larrea*, *Prosopis*, and *Acacia* are among the most important components of the lowland vegetation and are principally responsible for this similarity of aspect. Various other genera are shared by the two areas. Some notably desert-adapted genera are found in one area but not in the other (Morello, 1958; Raven, 1963; Axelrod, 1970).

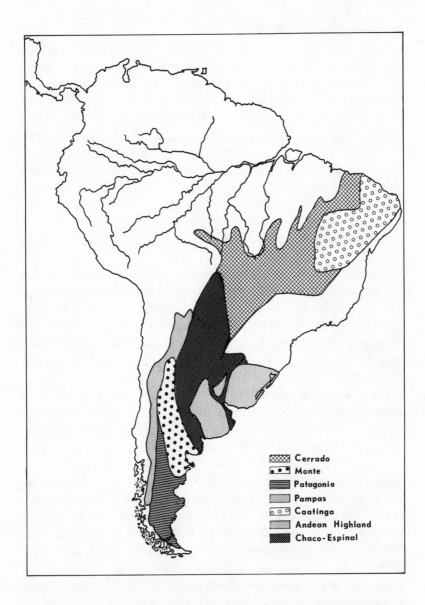

Fig. 8-1. Approximate distribution of xeric and subxeric areas in eastern and southern South America (adapted from Cabrera, 1953; Veloso, 1966; Sick, 1969).

Size of Area

The present areas of the Sonoran desert and the Monte are roughly similar in size. However, in considering the evolution of the desert-adapted fauna of the two continents, it is important to consider all contiguous desert areas. In this context the desertic areas of North America far exceed those that exist east of the Andes in South America. In South America there is only the Patagonian area with a cold desertic climate and the cold Andean Puna. In North America the addition of the Great Basin desert, the Mojave, and the Chihuahuan desert provides a much greater geographical expanse in which desert adaptations are favored.

Potential Sources of Stocks

The probability of any particular taxon of animal contributing to the fauna of either desert area obviously can be expected to decrease with the distance of that taxon's range from the desert area in question. This should be true not only because of the mere matter of distance but also because the more distant taxa would be expected to be adapted to the more distant and, hence, usually more different environments.

The nature of the adjacent ecological areas is, therefore, important to the process of evolution of the desert faunas. The Monte lies east of the Andean cordillera, which is a highly effective barrier to the interchange of lowland biota. To the south is the cold, desertic Patagonia, smaller in area than the Monte itself. To the east the Monte grades into the semiyaric thorn forest of the Chaco, which extends into Paraguay and Uruguay and merges into the Cerrado and Caatinga of Brazil. East of the Chaco are the pampa grasslands between roughly lat. 31° and 38°S (fig. 8-1). With the huge area of Chaco, Cerrado, and Caatinga to the east and northeast, and with the Chaco showing a strong gradient of decreasing moisture from east to west, we might expect this eastern area to be a likely source for the evolution of Monte species of terrestrial vertebrates.

The geographical relationship of the Sonoran desert to possible source areas for invading species is very different from that of the Monte. Mountains are to the west, but beyond that little similarity exists. For one thing, the Sonoran desert is part of a huge expanse of desertic areas that stretches over 3,000 km from the southern part of the Chihuahuan desert in Mexico to the northern tip of the Great Basin desert in Oregon. To the east of these deserts in the United States, beyond the Rocky Mountain chain, are the huge central grass-lands extending from the Gulf of Mexico into southern Canada. A similarity to the South American situation is seen, however, in the presence of a thorny vegetation type (the Mesquital), comparable to the Chaco, on the Gulf of Mexico lowlands of Tamaulipas and southern Texas. As in Argentina, a gradient of decreasing moisture exists westward from this Mesquital through the Chihuahuan desert and into the Sonoran desert. By contrast with the Monte, the Sonoran desert seems much more exposed to invasion by taxa which have adapted toward warm-xeric conditions in other contiguous areas.

Past Regional History

The present character of the two desert faunas obviously relates to the past histories of the two regions. For how long has there been selec-tion for a xeric-adapted fauna in each area? What have been the ef-fects on these faunas of secular climatic changes in the Tertiary and Pleistocene? These questions are difficult to answer with any great precision.

According to Axelrod (1948, p. 138, and other papers), "the pres-ent desert vegetation of the western United States, as typified by the floras of the Great Basin, Mohave and Sonoran deserts" is no older than middle Pliocene. Prior to the Oligocene, a Neotropical-Tertiary geoflora extended from southeastern Alaska and possibly Nova Scotia south into Patagonia (Axelrod, 1960) and began shrinking poleward as the continent became cooler and drier from the Oligo-cene onward. With respect to the Monte, Kusnezov (1951), as quoted by Morello (1958), believed that the Monte has existed without major change since "Eocene-Oligocene" times.

Arguments have been presented that there was a Gondwanaland

dry flora prior to the breakup of that land mass in the Cretaceous, which is represented today by xeric relicts in southern deserts (Axelrod, 1970). It seems then that selection for xeric adaptation has been going on in the southern continent and, from paleobotanical evidence, in North America as well (Axelrod, 1970, p. 310) for more than 100 million years. However, major climatic changes have occurred in the geographic areas now known as the Monte and the Sonoran desert. The present desert floras of these two areas are combinations of the old relicts and of types that have evolved as the continents dried and warmed from the Oligocene onward (Axelrod, 1970).

One of the unanswered questions is where the desert-adapted biotas were at times of full glaciations in the Pleistocene. Martin and Mehringer (1965, p. 439) have addressed this question with respect to North American deserts and have concluded that "Sonoran desert plants may have been hard pressed." The question is yet unanswered. The desert plants presumably retreated southward, but the degree of compression of their ranges is unknown. Doubt also exists whether the Monte biota could have remained where it now is at peaks of glaciation in the Southern Hemisphere (Simpson Vuilleumier, 1971).

The Anurans

The number of species of frogs is not greatly different for the two deserts, and, as might be expected, both faunas are relatively small. As we define the two faunas on the basis of present knowledge, the Sonoran desert fauna includes eleven species representing four families and four genera, while that of the Monte includes fourteen species representing three families and seven genera (table 8-1). (Definition of the Monte fauna is less certain and more arbitrary than that of the Sonoran because of scarcity of data. The listings of Monte and Chacoan species used here are based largely on data from Freiberg [1942], Cei [1955a, 1955b, 1959b, 1962], Reig and Cei [1963], and Barrio [1964a, 1964b, 1965a, 1965b, 1968] and on my own observations. Species recorded from Patquia in the province of La Rioja and from Alto Pencoso on the San Luis–Mendoza border [Cei, 1955a, 1955b] are included in the Monte fauna as here considered.)

Table 8-1. *Anuran Faunas: Monte of Argentina and Sonoran Desert of North America*

Sonoran	Monte
Pelobatidae	Ceratophrynidae
Scaphiopus couchi	*Ceratophrys ornata*
S. hammondi	*C. pierotti*
	Lepidobatrachus llanensis
	L. asper
Bufonidae	Bufonidae
Bufo woodhousei	*Bufo arenarum*
B. cognatus	
B. mazatlanensis	
B. retiformis	
B. punctatus	
B. alvarius	
B. microscaphus	
Hylidae	
Pternohyla fodiens	
Ranidae	Leptodactylidae
Rana sp.	*Odontophrynus occidentalis*
(*pipiens* gp.)	*O. americanus*
	Leptodactylus ocellatus
	L. bufonius
	L. prognathus
	L. mystaceus
	Pleurodema cinerea
	P. nebulosa
	Physalaemus biligonigerus

The composition of the two faunas is phylogenetically quite dissimilar. The Sonoran is dominated by members of the genus *Bufo* with seven species. The Monte fauna is dominated by leptodactylids with nine species distributed among four genera of that family.

Ecological similarities are evident between the two pelobatids (*Scaphiopus couchi* and *S. hammondi*) of the Sonoran fauna and the four ceratophrynids (*Ceratophrys ornata*, *C. pierotti*, *Lepidobatrachus asper*, and *L. llanensis*) of the Monte. The Sonoran has a single fossorial hylid (*Pternohyla fodiens*); I have found no evidence of a Monte hylid. However, a remarkably xeric-adapted hylid, *Phyllomedusa sauvagei*, extends at least into the dry Chaco (Shoemaker, Balding, and Ruibal, 1972); and, because of these adaptations, it would not be surprising to find it in the Monte. The canyons of the desert mountains of the Sonoran and Monte have a single species of *Hyla* of roughly the same size and similar habits. In Argentina it is *H. pulchella*; in the United States it is *H. arenicolor*. These are not included in our faunal listing for the two areas. The Sonoran has a ranid (*Rana* sp. [*pipiens* gp.]); the family has penetrated only the northern half of South America (with a single species) from old-world origins and via North America, so has had no opportunity to contribute to the Monte fauna.

The origins of the Monte anuran fauna seem relatively simple. This fauna is principally a depauperate Chacoan fauna (table 8-2). At least thirty-seven species of anurans are included in the Chacoan fauna. Every species in the Monte fauna also occurs in the Chaco. Nine of the fourteen Monte species have ranges that lie mostly within the combined Chaco-Monte. The Monte fauna thus represents that component of a biota which has had a long history of adaptation to xeric or subxeric conditions and is able to occupy the western, xeric end of a moisture gradient that extends from the Atlantic coast west to the base of the Andes. Two of the Monte species (*Leptodactylus mystaceus* and *L. ocellatus*) are wide-ranging tropical species that reach both the Monte and the Chaco from the north or east. We are treating *Odontophrynus occidentalis* as a sub-Andean species (Barrio, 1964a), but the genus has the Chaco-Monte distribution; and since this species reaches the Atlantic coast in Buenos Aires province, there is no certainty that it evolved in the Monte. *Pleurodema nebulosa* of the Monte is listed by Cei (1955b, p. 293) as "a characteristic cordilleran form"; and, as mapped by Barrio (1964b), its range barely enters the Chaco, although other members of the same species group occur in the dry Chaco. *Pleurodema cinerea* is treated

Table 8-2. *Comparison of Chaco and Monte Anuran Faunas*

Monte	Chaco
	Hypopachus mulleri
Ceratophrys ornata	*Ceratophrys ornata*
C. pierotti	*C. pierotti*
Lepidobatrachus llanensis	*Lepidobatrachus llanensis*
	L. laevis
L. asper	*L. asper*
Pleurodema nebulosa	*Pleurodema nebulosa*
	P. quayapae
	P. tucumana
P. cinerea	*P. cinerea*
Physalaemus biligonigerus	*Physalaemus biligonigerus*
	P. albonotatus
Leptodactylus ocellatus	*Leptodactylus ocellatus*
	L. chaquensis
L. bufonius	*L. bufonius*
L. prognathus	*L. prognathus*
L. mystaceus	*L. mystaceus*
	L. sibilator
	L. gracilis
	L. mystacinus
Odontophrynus occidentalis	*Odontophrynus occidentalis*
O. americanus	*O. americanus*
Bufo arenarum	*Bufo arenarum*
	B. paracnemis
	B. major
	B. fernandezae
	B. pygmaeus
	Melanophryniscus stelzneri
	Pseudis paradoxus
	Lysapsus limellus

Monte	Chaco
	Phyllomedusa sauvagei
	P. hypochondrialis
	Hyla pulchella
	H. trachythorax
	H. venulosa
	H. phrynoderma
	H. nasica

Note: All species listed for Monte occur also in Chaco.

by Gallardo (1966) as a member of his fauna "Subandina." The genus ranges north to Venezuela.

The Sonoran anurans seemingly have somewhat more diverse geographical origins than those of the Monte, and they have been more thoroughly studied. Most of the ranges can be interpreted as ones that have undergone varying degrees of expansion northward following full glacial displacement into Mexico (Blair, 1958, 1965). Several of these (*Scaphiopus couchi*, *S. hammondi*, and *Bufo punctatus*) have a main part of their range in the Chihuahuan desert (table 8-3). *Bufo cognatus* ranges far northward through the central grasslands to Canada. Three species extend into the Sonoran from the lowlands of western Mexico. One of these is the fossorial hylid *Pternohyla fodiens*. Another, *B. retiformis*, is one of a three-member species group that ranges from the Tamaulipan Mesquital westward through the Chihuahuan desert into the Sonoran. The third, *B. majallanorum*, is a member of a species group that is absent from the Chihuahuan desert but is represented in the Tamaulipan thorn scrub. *Bufo woodhousei* has an almost transcontinental range. *Rana* sp. is an undescribed member of the *pipiens* group.

Two desert-endemic species occur in the Sonoran. One is *Bufo alvarius*, which appears to be an old relict species without any close living relative. *B. microscaphus* occurs in disjunct populations in the Chihuahuan, Sonoran, and southern Great Basin deserts. These populations are clearly relicts from a Pleistocene moist phase extension of the eastern mesic-adapted *B. americanus* westward into the present desert areas (A. P. Blair, 1955; W. F. Blair, 1957).

Table 8-3. *Comparison of Anuran Faunas of Sonoran Desert with Those of Chihuahuan Desert and Tamaulipan Mesquital*

Sonoran Desert	Chihuahuan-Tamaulipan
	Rhinophrynus dorsalis
	Hypopachus cuneus
	Gastrophryne olivacea
Scaphiopus hammondi	*Scaphiopus hammondi*
	S. bombifrons
S. couchi	*S. couchi*
	S. holbrooki
	Leptodactylus labialis
	Hylactophryne augusti
	Syrrhopus marnocki
	S. campi
	Bufo speciosus
Bufo cognatus	*B. cognatus*
B. punctatus	*B. punctatus*
	B. debilis
	B. valliceps
B. woodhousei	*B. woodhousei*
B. retiformis	
B. mazatlanensis	
B. alvarius	
B. microscaphus	
	Hyla cinerea
	H. baudini
	Pseudacris clarki
	P. streckeri
	Acris crepitans
Pternohyla fodiens	
Rana sp. (*pipiens* gp.)	*Rana* sp. (*pipiens* gp.)
	R. catesbeiana

Desert Adaptedness

If taxonomic diversity is taken as a criterion, the Monte fauna presents an impressive picture of desert adaptation. The genera *Odontophrynus* and *Lepidobatrachus* are both xeric adapted and are endemic to the xeric and subxeric region encompassed in this discussion. Three of the four leptodactylid genera which occur in the Monte (*Leptodactylus*, *Pleurodema*, and *Physalaemus*) are characterized by the laying of eggs in foam nests, either on the surface of the water or in excavations on land. This specialization may have a number of advantages, but one of the important ones would be protection from desiccation (Heyer, 1969).

In North America the only genus that can be considered a desert-adapted genus is *Scaphiopus*. This genus has two distinct subgeneric lines which, based on the fossil record, apparently diverged in the Oligocene (Kluge, 1966). Each subgenus is represented by a species in the Sonoran desert. Origin of the genus through adaptation of forest-living ancestors to grassland in the early Tertiary has been suggested by Zweifel (1956). *Pternohyla* is a fossorial hylid that apparently evolved in the Pacific lowlands of Mexico "in response to the increased aridity during the Pleistocene" (Trueb, 1970, p. 698). The diversity of *Bufo* species (*B. mazatlanensis*, *B. cognatus*, *B. punctatus*, and *B. retiformis*) that represent subxeric- and xeric-adapted species groups and the old relict *B. alvarius* implies a long history of *Bufo* evolution in arid and semiarid southwestern North America. Nevertheless, the total anuran diversity of xeric-adapted taxa compares poorly with that in South America.

The greater taxonomic diversity of desert-adapted South American anurans may be attributed to the Gondwanaland origin (Reig, 1000, Casamiquela, 1961; Blair, 1973) of the anurans and the long history of anuran radiation on the southern continent. The taxonomic diversity of anurans in South America vastly exceeds that in North America, which has an attenuated anuran fauna that is a mix of old-world emigrants (Ranidae, possibly Microhylidae) and invaders from South America (Bufonidae, Hylidae, and Leptodactylidae). The drastic effects of Pleistocene glaciations on North American environments may also account for the relatively thin anuran fauna of this continent.

Mechanisms of Desert Adaptation

Limited availability of water to maintain tissue water in adults and un-predictability of rains to permit reproduction and completion of the lar-val stage are paramount problems of desert anurans. Enough is known about the ecology, behavior, and physiology of the anurans of the two deserts to indicate the principal kinds of mechanisms that have evolved in the two areas.

With respect to the first of these two problems, two major and quite different solutions are evident in both desert faunas. One is to avoid the major issue by becoming restricted to the vicinity of permanent water in the desert environment. The other is to become highly fos-sorial, to evolve mechanisms of extracting water from the soil, and to become capable of long periods of inactivity underground. In the Sonoran desert three of the eleven species fit the first category. The *Rana* species is largely restricted to the vicinity of water throughout its range to the east and is a member of the *R. pipiens* complex, which is essentially a littoral-adapted group. Ruibal (1962*b*) studied a desert population of these frogs in California and regards their winter breed-ing as an adaptation to avoid the desert's summer heat. The relict endemic *Bufo alvarius* is smooth skinned and semiaquatic (Steb-bins, 1951; my data). The relict populations of *B. microscaphus* oc-cur where there is permanent water as drainage from the mountains or as a result of irrigation. Man's activities in impounding water for ir-rigation must have been of major assistance to these species in invad-ing a desert region without having to cope with the major water prob-lems of desert life. *Bufo microscaphus*, for example, exists in areas that have been irrigated for thousands of years by prehistoric cultures and more recently by European man (Blair, 1955). One species in the Monte fauna is there by this same adaptive strategy. *Leptodactylus ocellatus* offers a striking parallel to the *Rana* species. Its existence in the provinces of Mendoza and San Juan is attributed to extensive ag-ricultural irrigation (Cei, 1955*a*). That a second species, *B. arenarum*, fits this category is suggested by Ruibal's (1962*a*, p. 134) statement that "this toad is found near permanent water and is very common around human habitations throughout Argentina." However, Cei (1959*a*) has shown experimentally that *B. arenarum* from the Monte

(Mendoza) survives desiccation more successfully than *B. arenarum* from the Chaco (Córdoba), which implies exposure and adaptation to more rigorously desertic conditions for the former.

Most of the anurans of both desert faunas utilize the strategy of subterranean life to avoid the moisture-sapping environment of the desert surface. In the Sonoran fauna the two species of *Scaphiopus* have received considerable study. One of these, *S. couchi*, appears to have the greatest capacity for desert existence. Mayhew (1962, p. 158) found this species in southern California at a place where as many as three years might pass without sufficient summer rainfall to "stimulate them to emerge, much less successfully reproduce."

Mayhew (1965) listed a series of presumed adaptations of this species to desert environment:

1. Selection of burial sites beneath dense vegetation where reduced insolation reaching the soil means lower soil temperatures and reduced evaporation from the soil
2. Retention by buried individuals of a cover of dried, dead skin, thus reducing water loss through the skin
3. Rapid development of larvae—ten days from fertilization through metamorphosis (reported also by Wasserman, 1957)

Physiological adaptations of *S. couchi* (McClanahan, 1964, 1967, 1972) include:

1. Storage of urea in body fluids to the extent that plasma osmotic concentration may double during hibernation
2. Muscles showing high tolerance to hypertonic urea solutions
3. Rate of production of urea a function of soil water potential
4. Fat utilization during hibernation
5. Ability to tolerate water loss of 48–58 percent of standard weight
6. Ability to store up to 30 percent of standard body weight as dilute urine to replace water lost from body fluids

The larvae of *S. couchi* are more tolerant of high temperatures than anurans from less-desertic environments, and tadpoles have been observed in nature at water temperatures of 39° to 40°C (Brown, 1969).

Scaphiopus hammondi, as studied by Ruibal, Tevis, and Roig (1969) in southeastern Arizona, shows a pattern of desert adaptation generally comparable to that of *S. couchi* but with some difference in details. These spadefoots burrow underground in September to depths of up to 91 cm and remain there until summer rains come some nine months later. The burrows are in open areas, not beneath dense vegetation as reported for *S. couchi* by Mayhew (1965). *S. hammondi* can effectively absorb soil water through the skin and has greater ability to absorb soil moisture "than that demonstrated for any other amphibian" (Ruibal et al., 1969, p. 571). During the rainy season of July–August, the *S. hammondi* burrows to depths of about 4 cm.

Larval adaptations of *S. hammondi* include rapid development and tolerance of high temperatures (Brown, 1967a, 1967b), paralleling the adaptations of *S. couchi*.

The adaptations of *Bufo* for life in the Sonoran desert are less well known than those of *Scaphiopus*. Four of the nonsemiaquatic species escape the rigors of the desert surface by going underground. *Bufo cognatus* and *B. woodhousei* have enlarged metatarsal tubercles or digging spades, as in *Scaphiopus*. In southeastern Arizona, *B. cognatus* was found buried at the same sites as *S. hammondi* but in lesser numbers (Ruibal et al., 1969). McClanahan (1964) found the muscles of *B. cognatus* comparable to those of *S. couchi* in tolerance to hypertonic urea solutions, a condition which he regarded as a fossorial-desert adaptation. *Bufo punctatus* has a flattened body and takes refuge under rocks. It has been reported from mammal (*Cynomys*) burrows (Stebbins, 1951). *Bufo punctatus* has the ability to take up water rapidly from slightly moist surfaces through specialization of the skin in the ventral pelvic region ("sitting spot"), which makes up about 10 percent of the surface area of the toad (McClanahan and Baldwin, 1969). *Bufo retiformis* belongs to the arid-adapted *debilis* group of small but very thick-skinned toads (Blair, 1970).

The Sonoran desert species of *Bufo* have not evolved the accelerated larval development that is characteristic of *Scaphiopus*. Zweifel (1968) determined developmental rates for three species of *Scaphiopus*, three species of *Bufo, Hyla arenicolor*, and *Rana* sp. (*pipiens* gp.) in southeastern Arizona. The eight species fell into three groups: most rapid, *Scaphiopus*; intermediate, *Bufo* and *Hyla*; slowest, *Rana*. In my laboratory (table 8-4) *B. punctatus* from central Arizona showed no acceleration of development over the same species from the extreme eastern part of the range in central Texas. *Bufo cognatus* closely paralleled *B. punctatus* in duration of the lar-

Table 8-4. *Duration of Larval Stage of Four of the Sonoran Desert Species of* Bufo

Species	Locality of Origin	Days from Fertilization to Metamorphosis		Lab Stock No.
		First	50%	
B. punctatus	Wimberley, Texas	27	32	B64–173
B. punctatus	Mesa, Arizona	27	36	B64–325
B. cognatus	Douglas, Arizona	28	35	B64–234
B. mazatlanensis	Mazatlan, Sinaloa × Ixtlan, Nayarit	20	26	B63–87
B. alvarius	Tucson × Mesa, Arizona	36	53	B65–271
B. alvarius	Mesa, Arizona	29	33	B64–361

Note: Observations in a laboratory environment at 80° F ± I.

val stage; *B. mazatlanensis* had a somewhat shorter larval life than these others; and *B. alvarius* spent a slightly longer period as tadpoles, but this could be accounted for by the fact that these are much larger toads. Overall, the impression is that these *Bufo* species have not shortened the larval stage as a desert adaptation. Tevis (1966) found that *B. punctatus* that were spawned in spring in Deep Canyon, California, required approximately two months for metamorphosis.

Developing eggs of *B. punctatus* and *B. cognatus* from Mesa, Arizona, were tested for temperature tolerances by Ballinger and Mc-Kinney (1966). Both of these desert species were limited by lower maxima than was *B. valliceps*, a nondesert toad, from Austin, Texas.

The fossorial anurans of the Monte are much less well known than those of the Sonoran desert. The ceratophrynids appear to be rather similar to *Scaphiopus* in their desert adaptations. Both species of *Lepidobatrachus* are reported to live buried (*viven enterrados*) and emerge after rains (Reig and Cei, 1963). *Lepidobatrachus llanensis* forms a cocoon made of many compacted dead cells of the stratum corneum when exposed to dry conditions (McClanahan, Shoemaker, and Ruibal, 1973). These anurans apparently live an aquatic exist-ence as long as the temporary rain pools exist, in which respect they differ from *Scaphiopus* species, which typically breed quickly and leave the water. The skin of *L. asper* is described (Reig and Cei, 1963) as thin in summer (when they are aquatic) and thicker and more granular in periods of drought. *Ceratophrys* reportedly uses the bur-rows of the viscacha (*Lagidium*), a large rodent (Cei, 1955*b*). How-ever, *C. ornata* does bury itself in the soil, and one was known to stay underground between four and five months and shed its skin after emerging (Marcos Freiberg, 1973, personal communication). *Ceratophrys pierotti* remains near the temporary pools in which it breeds for a considerable time after breeding (my observations). *Odontophrynus* at Buenos Aires makes shallow depressions and may sit in these with only the head showing (Marcos Freiberg, 1973, personal communication). *Leptodactylus bufonius* lives in dens or natural cavities or in viscacha burrows (Cei, 1949, 1955*b*). *Pleuro-dema nebulosa* is a fossorial species with metatarsal spade that spends a major portion of its lifetime living on land in burrows (Rui-bal, 1962*a*; Gallardo, 1965). *Bufo arenarum* "winters buried up to a meter in depth" (Gallardo, 1965, p. 67).

Phyllomedusa sauvagei of the dry Chaco, and possibly the Monte, has achieved a high level of xeric adaptation by excreting uric acid and by controlling water loss through the skin (Shoemaker et al., 1972). Rates of water loss in this arboreal, nonfossorial hylid are com-parable to those of desert lizards rather than to those of other anurans (Shoemaker et al., 1972).

Ruibal (1962*a*) studied the osmoregulation of six of the Chaco-Monte species and found that *P. nebulosa* is capable of producing urine that is hypotonic to the lymph and to the external medium, thus enabling it to store bladder water as a reserve against dehydration. The others, including *P. cinerea*, *L. asper*, and *B. arenarum* of what we are calling the Monte fauna, produced urine that was essentially isotonic to the lymph and the external medium.

Reproductive Adaptations

One of the major hazards of desert existence for an anuran population is the unpredictability of rainfall to provide breeding pools. Two alternative routes are available. One is to be an opportunistic breeder, spending long periods of time underground but responding quickly when suitable rainfall occurs. The alternative is to breed only in permanent water, with the time of breeding presumably set by such cues as temperature or possibly photoperiod. Both strategies are found among the Sonoran desert anurans.

The two *Scaphiopus* species are the epitome of the first of these adaptive routes. *Bufo cognatus*, *B. retiformis*, and *Pternohyla fodiens* are also opportunistic breeders (Lowe, 1964; my data). Two species, *B. punctatus* and *B. woodhousei*, are opportunistic breeders or not, depending on the population. Both are opportunistic in Texas. In the Great Basin desert of southwestern Utah, these two species along with all other local anurans (*B. microscaphus*, *S. intermontanus*, *Hyla arenicolor*, and *Rana* sp. [*pipiens* gp.]) breed without rainfall (Blair, 1955; my data). Peak breeding choruses of *B. punctatus* and *B. alvarius* were found in a stock pond near Scottsdale, Arizona, in the addition at any recent rain (Blair and Pettus 1954)

The Monte anurans, with the presumed exception of *Leptodactylus ocellatus*, appear to be opportunistic breeders (Cei, 1955a, 1955b; Reig and Cei, 1963; Gallardo, 1965; Barrio, 1964b, 1965a, 1965b). The apparent lesser development of the strategy of permanent water breeders could result from lesser knowledge of the behavior of the Monte anurans. However, the available evidence points to a real difference between the Monte and Sonoran desert faunas in degree of adoption of the habit of breeding in permanent water. *Leptodactylus ocellatus* of the Monte is ecologically equivalent to *R.* sp. (*pipiens* gp.) of the Sonoran desert; both are littoral adapted over a wide geographic range and have been able to penetrate their respective deserts by virtue of this adaptation where permanent water exists. There is no evidence that permanent water breeders are evolving from opportunistic breeders as in *B. punctatus*, *B. woodhousei*, and other North American desert species.

Foam Nests

One mechanism for desert adaptation, the foam nest, has been available for the evolution of the Monte fauna but not for the Sonoran desert fauna. Evolution of the foam-nesting habit has been discussed by various authors, especially Lutz (1947, 1948), Heyer (1969), and Martin (1967, 1970). The presumably more primitive pattern of floating the foam nest on the surface of the water is found among the Monte anurans in the genera *Physalaemus* and *Pleurodema* and in *Leptodactylus ocellatus*. The three other species of *Leptodactylus* in the Monte fauna lay their eggs in foam nests in burrows near water. These have aquatic larvae which are typically flooded out of the nests when pool levels rise with later rainfall. Heyer (1969) discussed advantages of the burrow nests over floating foam nests, among which the most important as adaptations to desert conditions are greater freedom from desiccation, and getting a head start on other breeders in the pool and thus being able to metamorphose earlier than others. Shoemaker and McClanahan (1973) investigated nitrogen excretion in the larvae of *L. bufonius* and found these larvae highly urotelic as an apparent adaptation to confinement in the foam-filled burrow versus the usual ammonotelism of anuran larvae.

Leptodactylids do reach the North American Mesquital (table 8-3), and one burrow-nesting species (*L. labialis*) reaches the southern tip of Texas. The other two genera both have direct, terrestrial development and hence would be unlikely candidates for desert adaptation. *Leptodactylus labialis* with a nesting pattern similar to that of *L. bufonius* would seem to be potential material for desert adaptation.

Cannibalism

An intriguing similarity between the two desert faunas is seen in the occurrence of cannibalism in both areas and in groups (ceratophrynids in South America, *Scaphiopus* in North America) that in other respects show rather similar patterns of desert adaptation.

In *S. bombifrons* and the closely related *S. hammondi*, some larvae have a beaked upper jaw and a corresponding notch in the lower as an apparent adaptation for carnivory (Bragg, 1946, 1950, 1956, 1961, 1964; Turner, 1952; Orton, 1954; Bragg and Bragg, 1959). The larvae of this type have been observed to be cannibalistic in *S. bombifrons* and suspected of being so in *S. hammondi* (Bragg, 1964). Cannibalism could be an important mechanism for concentrating food resources in a part of the population where these are limited and where there is a constant race against drying up of the breeding pool in the desert environment.

The ceratophrynids are much more cannibalistic than *Scaphiopus*. Both larvae and adults are carnivorous and cannibalistic (Cei, 1955*b*; Reig and Cei, 1963; my data). The head of the adult ceratophrynid is relatively large, with wide gape and with enlarged grabbing and holding teeth. Adult *Ceratophrys pierotti* are extremely voracious cannibals; one of these can quickly ingest another individual of its own body size.

Summary

The Monte of Argentina and the Sonoran desert of North America are compared with respect to their anuran faunas. Both deserts are roughly of similar size, but in North America there is a much greater extent of arid lands than in South America, with the Sonoran desert only a part of this expanse. Both deserts are at the dry end of moisture gradients that extend from thorn forest in the east to desert on the west.

Paleobotanical evidence suggests that xeric adaptation may have been occurring in South America prior to the breakup of Gondwanaland in the Cretaceous, while the North American deserts seem no older than middle Pliocene. Both desert systems must have been pressured and shifted during Pleistocene glacial maxima.

The anuran faunas of the two areas are similar in size, eleven species in the Sonoran desert, fourteen in the Monte. All anurans of the Monte occur also in the Chaco, and the fauna of the Monte is simply a depauperate Chacoan fauna. The origins of the Sonoran desert fauna are more diverse than this.

The Monte has the greatest taxonomic diversity, with seven genera versus four for the Sonoran desert. Two of the Monte genera (*Odontophrynus* and *Lepidobatrachus*) are truly desert and subxeric genera, but only one North American genus (*Scaphiopus*) fits this category. The presence of seven species of *Bufo* in the Sonoran desert implies a long history of desert adaptation by this genus in North America.

Mechanisms of desert adaptation are similar in the two areas. In each a littoral-adapted type (*Leptodactylus ocellatus* in the south, *Rana* sp. [*pipiens* gp.] in the north) has invaded the desert area by staying with permanent water. Additionally, the relict North American *B. alvarius* and *B. microscaphus* have followed the same strategy. Several of the North American species have abandoned opportunistic breeding in favor of breeding in permanent water, but no comparable trend is evident for the South American frogs. The most desert-adapted species in the North American desert is *Scaphiopus couchi*, which follows a pattern of highly fossorial life, opportunistic breeding with accelerated larval development, and physiological adaptations of adults to minimal water.

The ceratophrynids of the South American desert show parallel adaptations to those of *Scaphiopus*. In addition to other similarities, both groups employ some degree of cannibalism as an apparent adaptation to desert life.

References

Axelrod, D. I. 1948. Climate and evolution in western North America during middle Pliocene time. *Evolution* 2:127–144.

———. 1960. The evolution of flowering plants. In *Evolution after Darwin: Vol. 1 The evolution of life*, ed. S. Tax, pp. 227–305. Chicago: Univ. of Chicago Press.

———. 1970. Mesozoic paleogeography and early angiosperm history. *Bot. Rev.* 36:277–319.

Ballinger, R. E., and McKinney, C. O. 1966. Developmental temperature tolerance of certain anuran species. *J. exp. Zool.* 161:21–28.

Barbour, M. G., and Díaz, D. V. 1972. *Larrea* plant communities on bajada and moisture gradients in the United States and Argentina. *U.S./Intern. biol. Progn.: Origin and Structure of Ecosystems Tech. Rep.* 72–6:1–27.

Barrio, A. 1964*a*. Caracteres eto-ecológicos diferenciales entre *Odontophrynus americanus* (Dumeril et Bibron) y *O. occidentalis* (Berg) (Anura, Leptodactylidae). *Physis, B. Aires* 24:385–390.

———. 1964*b*. Especies crípticas del género *Pleurodema* que conviven en una misma área, identificados por el canto nupcial (Anura, Leptodactylidae). *Physis, B. Aires* 24:471–489.

———. 1965*a*. El género *Physalaemus* (Anura, Leptodactylidae) en la Argentina. *Physis, B. Aires* 25:421–448.

———. 1965*b*. Afinidades del canto nupcial de las especies cavicolas de género *Leptodactylus* (Anura, Leptodactylidae). *Physis, B. Aires* 25:401–410.

———. 1968. Revisión del género *Lepidobatrachus* Budgett (Anura, Ceratophrynidae). *Physis, B. Aires* 28:95–106.

Blair, A. P. 1955. Distribution, variation, and hybridization in a relict toad (*Bufo microscaphus*) in southwestern Utah. *Am. Mus. Novit.* 1722:1–38.

Blair, W. F. 1957. Structure of the call and relationships of *Bufo microscaphus* Cope. *Coepia* 1957:208–212.

———. 1958. Distributional patterns of vertebrates in the southern United States in relation to past and present environments. In *Zoogeography*, ed. C. L. Hubbs. *Publs Am. Ass. Advmt Sci.* 51:433–468.

———. 1965. Amphibian speciation. In *The Quaternary of the United States*, ed. H. E. Wright, Jr., and D. G. Frey, pp. 543–556. Princeton: Princeton Univ. Press.

———. 1970. Nichos zoológicos y la evolución paralela y convergente de los anfibios del Chaco y del Mesquital Norteamericano. *Acta zool. lilloana* 27:261–267.

———. 1973. Major problems in anuran evolution. In *Evolutionary biology of the anurans: Contemporary research on major problems*, ed. J. L. Vial, pp. 1–8. Columbia: Univ. of Mo. Press.

Blair, W. F., and Pettus, D. 1954. The mating call and its significance in the Colorado River toad (*Bufo alvarius* Girard). *Tex. J. Sci.* 6:72–77.

Bragg, A. N. 1946. Aggregation with cannibalism in tadpoles of

Scaphiopus bombifrons with some general remarks on the probable evolutionary significance of such phenomena. *Herpetologica* 3:89–98.

———. 1950. Observations on *Scaphiopus*, 1949 (Salientia: Scaphiopodidae). *Wasmann J. Biol.* 8:221–228.

———. 1956. Dimorphism and cannibalism in tadpoles of *Scaphiopus bombifrons* (Amphibia, Salientia). *SWest. Nat.* 1:105–108.

———. 1961. A theory of the origin of spade-footed toads deduced principally by a study of their habits. *Anim. Behav.* 9:178–186.

———. 1964. Further study of predation and cannibalism in spadefoot tadpoles. *Herpetologica* 20:17–24.

Bragg, A. N., and Bragg, W. N. 1959. Variations in the mouth parts in tadpoles of *Scaphiopus* (*Spea*) *bombifrons* Cope (Amphibia: Salientia). *SWest. Nat.* 3:55–69.

Brown, H. A. 1967*a*. High temperature tolerance of the eggs of a desert anuran, *Scaphiopus hammondi*. *Copeia* 1967:365–370.

———. 1967*b*. Embryonic temperature adaptations and genetic compatibility in two allopatric populations of the spadefoot toad, *Scaphiopus hammondi*. *Evolution* 21:742–761.

———. 1969. The heat resistance of some anuran tadpoles (Hylidae and Pelobatidae). *Copeia* 1969:138–147.

Cabrera, A. L. 1953. Esquema fitogeográfico de la República Argentina. *Revta Mus. La Plata (Nueva Serie), Bot.* 8:87–168.

Casamiquela, R. M. 1961. Un pipoideo fósil de Patagonia. *Revta Mus. La Plata Sec. Paleont. (Nueva Serie)* 4:71–123.

Cei, J. M. 1949. Costumbres nupciales y reproducción de un batracio caracteristico chaqueño (*Leptodactylus bufonius*). *Acta zool. lilloana* 8:105–110.

———. 1955*a*. Notas batracológicas y biogeográficas Argentinas, I–IV. *An. Dep. Invest. cient., Univ. nac. Cuyo.* 2(2):1–11.

———. 1955*b*. Chacoan batrachians in central Argentina. *Copeia* 1955:291–293.

———. 1959*a*. Ecological and physiological observations on polymorphic populations of the toad *Bufo arenarum* Hensel, from Argentina. *Evolution* 13:532–536.

———. 1959*b*. Hallazgos hepetológicos y ampliación de la distribución geográfica de las especies Argentinas. *Actas Trab. Primer Congr. Sudamericano Zool.* 1:209–210.

———. 1962. Mapa preliminar de la distribución continental de las "sibling species" del grupo *ocellatus* (género *Leptodactylus*). *Revta Soc. argent. Biol.* 38:258–265.

Freiberg, M. A. 1942. Enumeración sistemática y distribución geográfica de los batracios Argentinos. *Physis, B. Aires* 19:219–240.

Gallardo, J. M. 1965. Consideraciones zoogeográficas y ecológicas sobre los anfibios de la provincia de La Pampa Argentina. *Revta Mus. argent. Cienc. nat. Bernardino Rivadavia Inst. nac. Invest. Cienc. nat. Ecol.* 1:56–78.

―――. 1966. Zoogeografía de los anfibios chaqueños. *Physis, B. Aires* 26:67–81.

Heyer, W. R. 1969. The adaptive ecology of the species groups of the genus *Leptodactylus* (Amphibia, Leptodactylidae). *Evolution* 23:421–428.

Kluge, A. G. 1966. A new pelobatine frog from the lower Miocene of South Dakota with a discussion of the evolution of the *Scaphiopus-Spea* complex. *Contr. Sci.* 113:1–26.

Kusnezov, N. 1951. *La edad geológica del régimen árido en la Argentina ségun los datos biológicos*. Geográfica una et varia, *Publnes esp. Inst. Estud. geogr., Tucumán* 2:133–146.

Lowe, C. H., ed. 1964. *The vertebrates of Arizona*. Tucson: Univ. of Ariz. Press.

Lutz, B. 1947. Trends toward non-aquatic and direct development in frogs. *Copeia* 1947:242–252.

―――. 1948. Ontogenetic evolution in frogs. *Evolution* 2:29–39.

McClanahan, L. J. 1964. Osmotic tolerance of the muscles of two desert-inhabiting toads, *Bufo cognatus* and *Scaphiopus couchi*. *Comp. Biochem. Physiol.* 12:501–508.

―――. 1967. Adaptations of the spadefoot toad, *Scaphiopus couchi*, to desert environments. *Comp. Biochem. Physiol.* 20:73–99.

―――. 1972. Changes in body fluids of burrowed spadefoot toads as a function of soil potential. *Copeia* 1972:209–216.

McClanahan, L. J., and Baldwin, R. 1969. Rate of water uptake through the integument of the desert toad, *Bufo punctatus*. *Comp. Biochem. Physiol.* 29:381–389.

McClanahan, L. J.; Shoemaker, V. H.; and Ruibal, R. 1973. Evaporative water loss in a cocoon-forming South American anuran. Abstract of paper given at 53d Annual Meeting of American Society of Ichthyologists and Herpetologists, at San José, Costa Rica.

Martin, A. A. 1967. Australian anuran life histories: Some evolutionary and ecological aspects. In *Australian inland waters and their fauna*, ed. A. H. Weatherley, pp. 175–191. Canberra: Aust. Nat. Univ. Press.

————. 1970. Parallel evolution in the adaptive ecology of Leptodactylid frogs in South America and Australia. *Evolution* 24:643–644.

Martin, P. S., and Mehringer, P. J., Jr. 1965. Pleistocene pollen analysis and biogeography of the southwest. In *The Quaternary of the United States*, ed. H. W. Wright, Jr., and D. G. Frey, pp. 433–451. Princeton: Princeton Univ. Press.

Mayhew, W. W. 1962. *Scaphiopus couchi* in California's Colorado Desert. *Herpetologica* 18:153–161.

————. 1965. Adaptations of the amphibian, *Scaphiopus couchi*, to desert conditions. *Am. Midl. Nat.* 74:95–109.

Morello, J. 1958. La provincia fitogeográfica del Monte. *Op. lilloana* 2:1–155.

Orton, G. L. 1954. Dimorphism in larval mouthparts in spadefoot toads of the *Scaphiopus hammondi* group. *Copeia* 1954:97–100.

Raven, P. H. 1963. Amphitropical relationships in the floras of North and South America. *Q. Rev. Biol.* 38:141–177.

Reig, O. A. 1960. Lineamentos generales de la historia zoogeográfica de los anuros. *Actas Trab. Primer Congr. Sudamericano Zool.* 1:271–278.

Reig, O. A., and Cei, J. M. 1963. Elucidación morfológico-estadística de las entidades del género *Lepidobatrachus* Budgett (Anura, Ceratophrynidae) con consideraciones sobre la extensión del distrito chaqueño del dominio zoogeográfico subtropical. *Physis, B. Aires* 24:181–204.

Ruibal, R. 1962a. Osmoregulation in amphibians from heterosaline habitats. *Physiol. Zoöl.* 35:133–147.

————. 1962b. The ecology and genetics of a desert population of *Rana pipiens*. *Copeia* 1962:189–195.

Ruibal, R.; Tevis, L., Jr.; and Roig, V. 1969. The terrestrial ecology of the spadefoot toad *Scaphiopus hammondi*. *Copeia* 1969:571–584.

Shelford, V. E. 1963. *The ecology of North America*. Urbana: Univ. of Ill. Press.

Shoemaker, V. H., and McClanahan, L. J. 1973. Nitrogen excretion in the larvae of the land-nesting frog (*Leptodactylus bufonius*). *Comp. Biochem. Physiol.* 44A:1149–1156.

Shoemaker, V. H.; Balding, D.; and Ruibal, R. 1972. Uricotelism and low evaporative water loss in a South American frog. *Science, N.Y.* 175:1018–1020.

Sick, W. D. 1969. Geographical substance. In *Biogeography and ecology in South America*, ed. E. J. Fittkau, J. Illies, H. Klinge, G. H. Schwabe, and H. Sioli, 2: 449–474. The Hague: Dr. W. Junk.

Simpson Vuilleumier, B. 1971. Pleistocene changes in the fauna and flora of South America. *Science, N.Y.* 173:771–780.

Solbrig, O. T. 1975. The origin and floristic affinities of the South American temperate desert and semidesert regions. In *Evolution of desert biota*, ed. D. W. Goodall. Austin: Univ. of Texas Press.

Stebbins, R. C. 1951. *Amphibians of western North America*. Berkeley and Los Angeles: Univ. of Calif. Press.

Tevis, L., Jr. 1966. Unsuccessful breeding by desert toads (*Bufo punctatus*) at the limit of their ecological tolerance. *Ecology* 47: 766–775.

Trueb, L. 1970. Evolutionary relationships of casque-headed tree frogs with coossified skulls (family Hylidae). *Univ. Kans. Publs Mus. nat. Hist.* 18:547–716.

Turner, F. B. 1952. The mouth parts of tadpoles of the spadefoot toad, *Scaphiopus hammondi*. *Copeia* 1952:172–175.

Veloso, H. P. 1966. *Atlas florestal do Brasil*. Rio de Janeiro—Guanabara: Ministerio da Agricultura.

Wasserman, A. O. 1957. Factors affecting interbreeding in sympatric species of spadefoots (genus *Scaphiopus*). *Evolution* 11:321–338.

Zweifel, R. G. 1956. Two pelobatid frogs from the Tertiary of North America and their relationships to fossil and recent forms. *Am. Mus. Novit.* 1762:1–45.

———. 1968. Reproductive biology of anurans of the arid southwest, with emphasis on adaptation of embryos to temperature. *Bull. Am. Mus. nat. Hist.* 140:1–64.

Notes on the Contributors

John S. Beard was born in England and educated at Oxford University. After graduation he spent nine years in the Colonial Forest Service in the Caribbean and during this period studied for the degree of D.Phil. at Oxford. Soon after the end of the Second World War, he went to South Africa, where he was engaged in research on crop improvement in the wattle industry. In 1961 he was appointed director of King's Park in Perth, Western Australia, where, among other things, he was responsible for establishing a botanical garden. Ten years later he became director of the National Herbarium of New South Wales, a post from which he has recently retired. He is now devoting much of his time to the preparation of a series of detailed maps of Australian vegetation.

Dr. Beard has published books and papers on the vegetation of tropical America and has wide interests in plant ecology, biogeography, and systematics.

W. Frank Blair is professor of zoology at the University of Texas at Austin. He was born in Dayton, Texas. His first degree was taken at the University of Tulsa; he was awarded the M.S. at the University of Florida, and the Ph.D. at the University of Michigan. After eight years as a research associate there, he moved to a faculty position at the University of Texas, where he has been ever since.

At the inception of the International Biological Program, Dr. Blair became director of the Origin and Structure of Ecosystems section and, soon afterward, national chairman for the whole program. He also served as vice-president of the IBP on the international scale.

He has been involved in a wide range of personal research on vertebrate ecology and has worked extensively in Latin America as well as the United States. He is senior author of *Vertebrates of the United States* and has edited *Evolution in the Genus "Bufo"*—a subject on which much of his most recent research has concentrated.

David W. Goodall is Senior Principal Research Scientist at CSIRO Division of Land Resources Management, Canberra, Australia. Born and brought up in England, he studied at the University of London where he was awarded the Ph.D. degree, and, after a period of research in what is now Ghana, took up residence in Australia, of which country he is a citizen. He was awarded the D.Sc. degree of Melbourne University in 1953. He came to the United States in 1967 and the following year was invited to become director of the Desert Biome section of the International Biological Program then getting under way. This position he continued to hold until the end of 1973. During most of this period, and until the end of 1974, he held a position as professor of systems ecology at Utah State University.

His main research interests were initially in plant physiology, particularly in its application to agriculture and horticulture; but later he shifted his interest to plant ecology, especially statistical aspects of the subject, and to systems ecology.

Bobbi S. Low is associate professor of resource ecology at the University of Michigan. She was born in Kentucky and took her first degree at the University of Louisville and her doctorate at the University of Texas at Austin. After postdoctoral work at the University of British Columbia, she spent three years as a Research Fellow at Alice Springs, Australia, and returned to the United States in 1972.

Her main research interests have been in evolutionary ecology and in ecology of vertebrates in arid areas, both in the United States and in Australia.

James A. MacMahon is professor of biology at Utah State University and assistant director of the Desert Biome section of the International Biological Program. He was born in Dayton, Ohio, and took his first degree at Michigan State University and his doctorate at Notre Dame University, Indiana, in 1963. He then was appointed to a professorial position at the University of Dayton, and in 1971 he moved to Utah.

Though much of his research has been devoted to reptiles and Amphibia, he has also been concerned with plants, mammals, and invertebrates. In all these groups of organisms, he has mainly been interested in their ecology, particularly at community level, in relation to the arid-land environment.

A. R. Main is professor of zoology at the University of Western Australia. He was born in Perth; after military service during the Second World War, he returned there to take a first degree and then a doctorate at the University of Western Australia. He is a Fellow of the Australian Academy of Science.

He has done extensive research on the ecology of mammals and Amphibia in the Australian deserts, and he and his students have published numerous papers on the subject.

Guillermo Sarmiento is associate professor in the Faculty of Science, Universidad de Los Andes, Mérida, Venezuela. He was born in Mendoza, Argentina, and was educated at the University of Buenos Aires, where he was awarded a doctorate in 1965. He was appointed assistant professor and moved to Venezuela two years later.

His main research interests have been in tropical plant ecology, particularly as applied to savannah and to the vegetation of arid lands.

Otto T. Solbrig was born in Buenos Aires, Argentina. He took his first degree at the Universidad de La Plata and his Ph.D. at the University of California, Berkeley, in 1959. He worked at the Gray Herbarium, Harvard University, for seven years (during which period he became a U.S. citizen); after a period as professor of botany at the University of Michigan, he returned to Harvard University in 1969 as professor of biology and chairman of the Sub-Department of Organismic and Evolutionary Biology. Within the U.S. contribution to the International Biological Program, he served as director of the Desert Scrub subprogram of the Origin and Structure of Ecosystems section.

He has wide field experience in various parts of Latin America as well as in the United States. His main research interests have been in plant biosystematics, biogeography, and population biology.

Index